CONTEMPORARY INDUSTRIAL TEACHING

Solving everyday problems

by

RONALD J. BAIRD

Professor of Industrial Education
Eastern Michigan University
Ypsilanti, Michigan

South Holland, Ill.
THE GOODHEART-WILLCOX CO., Inc.
Publishers

INTRODUCTION

The outstanding teacher of industrial education must possess qualities of resourcefulness, leadership, and a command of technical and human learning relationships. It is for this purpose that CONTEMPORARY INDUSTRIAL TEACHING is presented. The complexity of our industrial world and the increasing depth of content in technical courses makes it necessary for the industrial education teacher to comprehend the total teaching-learning situation. This book provides an opportunity to study the essentials that are so vital for efficient instruction.

Careful consideration has been given to organizational content so that it may serve the needs of those preparing for or engaged in teaching the variety of technical subjects being offered in our schools. Emphasis is given to those qualities that make teaching and learning an exciting experience.

The book presents in logical sequence the responsibilities of the teacher in course planning, content selection, and the development of learning activities. These are followed by laboratory management, equipment selection, and safety. The final chapters are devoted to the teaching process... student motivation, presentations, instructional media and methodology, evaluation, and successful teacher-student relationships.

CONTEMPORARY INDUSTRIAL TEACHING is the result of extensive analysis of successful teachers over many years. It is the experience of numerous industrial education teachers brought together for those who seek the best in learning how to teach.

It is intended for college students preparing to teach industrial education in junior and senior high schools, vocational schools, community colleges, and industrial training programs. This book should also be of value to experienced teachers, administrators, and supervisors of industrial education programs.

CONTENTS

Chapter 1
THE INDUSTRIAL EDUCATION TEACHER

Teaching in the field of industrial education opens the door to many opportunities. The world of industry and the industrial education teacher work hand in hand to provide educational opportunities for persons in all walks of life.

Industry touches all people in their daily lives as they work, read, take care of their home, relax, or play. As a part of this industrial-social complex, the industrial education teacher has a multitude of responsibilities in developing "what it takes" to do an effective teaching job.

Terminology used concerning those who teach technical or industrial subjects is sometimes confusing. To argue the point of definitions would prove fruitless, for discussions of this nature have been going on since manual training was introduced into the schools. Your total understanding of the field of teaching about industry is our main concern. In this text the following terminology will be used to distinguish one type of teaching program from another:

INDUSTRIAL EDUCATION - A generic term used to encompass all types of education dealing with industry and technology in our society. See chart.

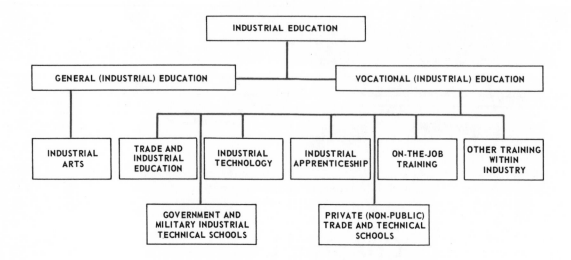

VOCATIONAL EDUCATION - The broad field of study designed to develop skills, attitudes, understandings, abilities, work habits and appreciations, including information and knowledge needed by workers to enter and successfully progress in employment on a productive basis. It is an integral part of the total program of education and contributes to the development of good citizens through the education of their social, cultural, and economic competencies. In this sense it includes preparation for any occupational endeavor, such as, home economics, trade and industrial education, distributive education, business education, and agricultural education.

VOCATIONAL INDUSTRIAL EDUCATION - Programs encompassing more specific objectives toward competencies necessary for entrance into a predetermined industrial, trade, or technical occupation. The vocational track in a comprehensive high school and extension classes for upgrading workers are typical examples.

TRADE AND INDUSTRIAL EDUCATION - Education designed to provide the necessary training for semiskilled or skilled craftsmen in processing, construction, production, assembling, servicing, or maintenance. The program usually consists of one-half day in the laboratory and one-half day in related general studies in the secondary school. Evening and on-the-job programs are also available as well as in-service training.

INDUSTRIAL ARTS - Phases of the general educational program which deal with contemporary American industry. This includes the study of industrial organization, materials, processes, products, and occupations - - and problems involving man and his technological society.

TECHNICAL EDUCATION - Educational programs planned for those who desire to earn a living in an occupation in which success is largely dependent upon technical information and understanding of the laws of science and technology as applied to design, manufacturing, production, distribution, and service. The program is generally post-high school and provides a background for employees called "technicians."

Opportunities for you as an industrial education teacher are almost unlimited. As you pursue your educational teacher preparation program, you should become acquainted with contemporary industrial education programs offered by junior and senior high schools, vocational and technical schools, community colleges, technical institutes, colleges, and universities. You will have an opportunity to select a specialization or technical area of interest in which you feel you can excel and make an outstanding contribution to society. Schools and institutions must have effective educational programs to carry out the mission of a well educated technically oriented society and a competent work force. You are the channel through which this mission may be accomplished.

Teaching industrial education is far more than "showing someone how to make something." This requires resourceful teachers who understand the problems of learning, the organization and function of contemporary American industry, the technical skills and "know how" of teaching, and the ability to relate to students. It is toward this goal that you should direct your thinking as you prepare to be a teacher of industrial education.

Teacher Qualifications and Education

"What makes an outstanding industrial education teacher?" Answering this question is a difficult task. Research has been done on this for years and few specific answers can be given. However, here's a brief list of desirable qualities:

RESOURCEFULNESS - DESIRE TO BE HELPFUL
PERSONALITY - EDUCATION - PATIENCE
CONFIDENCE - ENTHUSIASM
SINCERITY - LEADERSHIP

Four you should be particularly conscious of are: ENTHUSIASM, PATIENCE, RESOURCEFULNESS, SINCERITY.

The teacher who is enthusiastic and excited about his subject matter and teaching methods will usually convey these traits to the student. On the other hand, if the teacher gives the impression that the content is dull, the students will probably be so impressed. Observe the good teacher and take note of his enthusiasm during the teaching process.

Patience is like enthusiasm. You will find that all students do not progress, develop skills, or understand concepts as readily as you may expect. Give them a chance. The reward in observing a student finally master a complex mental or manipulative skill is well worth extra effort and waiting.

Resourcefulness; perhaps the greatest asset the outstanding industrial education teacher may attain. You will observe this quality in many good teachers. This refers to the teacher's ability to adapt to various learning situations, make innovations in teaching presentations, revise learning activities that "did not work," and to have at his fingertips sources of information which will continually upgrade his teaching. The resourceful teacher is constantly on the move. He does not rely on a folder of old notes or textbook as his teaching kit. He will be continually visiting industry, reading the latest technical and professional literature in his field, revising his laboratory, and preparing new teaching media as the daily problems of the teaching-learning situation unfold. Watch for this type of teacher as you go through your program. He does not just teach, he gives students an opportunity to learn - - there is quite a difference.

The successful teacher is a sincere teacher. He is proud of watching his students grow in their confidence in his ability. Without such confidence, the student will lose respect for the teacher and interest in the course. Sincerity can easily be detected by the student. If he knows you're not well prepared, bluffing, or providing "busy work" activities, he probably won't tell you but he will probably let others know. Sincerity also re-

lates to an unknown quantity on the part of the student. Not knowing at the time what should constitute the course content, he may appear satisfied. However, the common remark is often heard even years later; "I did not get what I really needed out of that course." Was the teacher sincere enough to see to it that he provided his best in preparation and teaching? Sincerity will be as rewarding to you as it is to the student.

The educational requirements for an industrial education teacher are, in part, dependent upon his goals. A bachelor's degree with a major in industrial education is the normal educational background for teaching industrial education in the public schools. You should constantly be concerned with the type of teaching position you would like to obtain and your long range goals. These plans will be subject to change, but they will serve as a guide to getting you where you would like to go. The plans are somewhat like scheduling a trip to a certain destination. You get out the maps and plot your route. There are many alternatives and considerations - - your method of transportation, finances, time alotted, speed, etc. Your educational trip also involves many considerations. Plan your goals, educational requirements necessary to get you there, and the best route to follow. Industrial experience, a master's degree, or a doctorate may be necessary requirements. Check with your advisor and other instructors to gain the benefit of their experience in planning educational requirements necessary to meet your personal goals.

Your Experience and Specialization

There are few industrial education teachers who can be classified as, "jack of all trades." Industry is far too complex and specialized for any one individual to be an expert in all phases of materials and production. It will therefore be necessary during your teacher preparation program to select an area or field of technical specialization in which you can develop your abilities to the maximum degree. Such areas as graphic arts,

electricity-electronics, plastics, metals, or drafting illustrate but a few. Some teachers will become highly specialized in a particular field, such as welding, while others may develop their skills in broad areas in preparation to teach manufacturing, production, or other segments of American industry. The emphasis given to any technical area is dependent upon the teaching requirements of the position one expects to hold.

It may be difficult for you to make a choice among technical areas as you begin your program. However, most colleges and universities have a basic core of study in industrial education which will allow you to explore and investigate the broad spectrum of industrial teaching pursuits. It is advisable for you to take advantage of the opportunity to evaluate the many choices you have for technical specialization.

Your past experiences may be a factor in technical area selection. You may have had some occupational experience which prompted you to enter the program. Your choice may already be well defined. For others, experiences in high school industrial education courses may have provided some interests and insights which will be helpful in making a choice. In any case, your keen interest in the content area in which you plan to teach will be a major factor in being a successful teacher. Continually develop this interest and you will be amazed at the satisfaction you receive as you plan and prepare for your teaching career.

Your Desire and Ability to Develop

How successful you are as an industrial education teacher depends to a considerable extent on your desire, and ability to develop as you progress through your program. Good teachers are not born with innate abilities. They are educated through the complex process of studying the nature of the teaching-learning process. Often you will hear, "the craftsman is the best teacher" and "you teach the way you were taught." These positions are debatable, yet little will be gained by extensive discussion. A balance between your technical skills and knowledge and your ability to be an outstanding teacher must be interwoven. Each depends heavily upon the other. You will probably teach the way you were taught, at least at the beginning.

Your desire and ability to develop will give you the background needed to move ahead. If a teacher were to state, "I've had ten years of teaching experience," we would want to qualify the statement. It may well be true. However, he may have had one year of teaching experience ten times. If he has not had the ability and desire to grow responsibly, he has not done the job required of a good teacher.

Your Professional Image

As an industrial education teacher you should take a great deal of pride in your professional image. This does not mean that you should be a "show off." What it really indicates is that you should develop those qualities that others admire in any professional person. When someone asks a person what his occupation is and the person replies, "I'm just a shop teacher," it is quite easy to imagine the image he has revealed. On the other hand, if he replied, "I teach industrial education," he is implying quite a different image.

Take advantage of the opportunities to present yourself as a professional person. Four broad guidelines may serve to assist you in this respect:

Dress neatly and appropriately for the situation at hand.

Learn to communicate with confidence on technical and professional topics.

Be just as active and enthusiastic outside your laboratory as within.

Take on leadership responsibilities in all phases of your educational system.

Your professional image reflects your occupation. Just as a neurosurgeon would scrub-up after an operation to attend a staff' meeting, you should wash the chalk off your hands, exchange your lab coat for a suit coat, and head for a faculty conference. Your course content may involve electrons, tool steels, or adhesives, but your professional image involves you. Prepare, and make it good.

Your Obligations to the Profession

A career in industrial education teaching requires you to meet a number of obligations to the profession, if you are to be successful. You will have responsibilities within your local school, and you will also have professional responsibilities to your community, state, and national organizations. You should pride yourself in being a member of a team of thousands of professional industrial education teachers who are responsible for the technical educational needs of our industrial society.

Teaching is a highly specialized and demanding profession. Industrial education teaching is particularly demanding. Besides your own personal growth, you will have professional obligations to your students, your total school program, your fellow teachers, and the associations to which you should give your support. Concentrate on being a professional teacher; it is rewarding.

Personal Obligations

Your personal obligations are numerous. It's much like the saying "no one is a great teacher but he is always becoming one." The task of being aware of your personal obligations to the profession is of major importance.

Look at three broad areas of personal behavior that should become part of your professional responsibility:

Maintain a good working relationship with your fellow teachers and you will receive their support and cooperation.

Develop a professional attitude toward your responsibilities.

Promote yourself as an example of character and technical "know how" that will inspire your students.

The teacher who is respected by his students will set a pattern for them to follow. If he has pride in his professional attitude, students will be inspired to do the same. The whole atmosphere of learning revolves around the personal obligations you have set for yourself.

Your Relationship to the Student

Most students will tend to emulate a good teacher and identify themselves with the character and skills that he possesses. As you grow in your study and understanding of teaching fundamentals, you will better understand the importance of teacher-student relationships. Think back to teachers you have had who stimulated you in a particular course. What professional qualities did they display that made you enjoy learning? Perhaps the following list will include many you should consider in your relationships with students:

a. Show a genuine interest in the problems and development of each student.
b. Be friendly and happy in your teaching; the dull teacher usually has dull students.
c. Be patient when students do not learn up to your expectations.
d. Make learning exciting by relating the content and yourself to the student.
e. Praise students for their accomplishments; you may not be their best buddy, but you may be their best teacher.
f. Let the student know that you do not have all the answers; that some of his may be the most important ones.

The importance of good teacher-student relationships cannot be overemphasized. As an industrial education teacher, it is one of your major professional obligations. The development of competent personnel for our industrial society will depend heavily on how well our industrial education teachers inspire students in the classroom and laboratory.

Supporting Your School Programs

As a professional teacher you also have an obligation to support your total school program. An institution will only be as good as those who administer and plan for learning. You are not an isolated teacher, unrelated to the other phases of educational, economic, and social development of the student. It is your responsibility as a professional educator to be a member of the "team" that is providing educational opportunities for all students. Some of the main concerns of total school support of which you should constantly be aware are:

a. Enter into and work on total school curriculum development.
b. Cooperate with other faculty on all school related problems.
c. Keep abreast of all new school innovations, available materials, and current philosophy.
d. Keep the administration and other faculty well informed of your progress and contributions in your field.
e. Accept and carry out assignments from the administration, committees and other faculty groups.

Only if the total faculty and administration cooperate in educational endeavors will the best education for students be provided.

You and Professional Associations

As a teacher of technical and industrial subjects you will have two major responsibilities in relation to professional associations. First, you should become a supporting member of those associations dealing with education and industrial education.

These would include such organizations as the American Vocational Association, the National Education Association, the American Industrial Arts Association at the national level, and similar groups at the state and regional level. Your participation in associations of this nature will not only upgrade the profession but also give you an opportunity to better understand and and keep informed as to the status of industrial education nationwide. You should also keep in mind what contributions you can make to professional associations as a part of your teaching career. You may want to write articles for publication in journals of various associations, take responsibilities on committees, hold office, or participate in convention programs. The total growth and advancement of contemporary programs in industrial education requires that all teachers, with varying interests and abilities, contribute to their professional organizations.

Secondly, and more specifically, you should take every opportunity to join and contribute to those professional associations in your technical area of specialization. These are usually industrial associations concerned with particular areas of manufacturing. They welcome the membership of educators. This is possibly the best way in which you can keep "on top" of the technical advancements in your field. Such associations as the National Electronics Teachers Service, International Graphic Arts Association, Society of the Plastics Industry, American Society for Metals, and the Forest Products Industry provide publications and literature which can give you probably the best in-service education program you can obtain. For example, a subscription to Modern Plastics Magazine, which includes the yearly Modern Plastics Encyclopedia, provides the plastics technology teacher with the latest developments in the plastics industry which he can incorporate into his course content at any level.

Teacher and professional associations provide a two-way road along which up-to-date information may be sent "special delivery" to the inquiring minds of the students.

Review Questions - Chapter 1

1. What does the term resourcefulness refer to as one of the qualities of a good teacher?
2. What is the difference between technical education and vocational education?
3. What opportunities are available to those preparing to teach in the field of industrial education?
4. What factors should you take into account as you pursue your goals in preparation to become an industrial education teacher?
5. How can you help yourself decide in which technical area of industrial education you may want to specialize?
6. What four guidelines are important in presenting yourself as a professional teacher?
7. What are a teacher's responsibilities in terms of professional associations in industrial education?
8. What are some of the responsibilities the industrial education teacher should assume concerning support of the total school program?
9. How does enthusiasm relate to good teaching?
10. Why is a good student-teacher relationship so important for potential learning situations?

Suggested Activities

1. Make a list of the qualities you personally feel are the most important for an industrial education teacher to possess.
2. Using reference literature, look up three modern definitions each for technical education, industrial arts, and vocational education for use in class discussion.
3. Prepare a chart indicating the organizational breakdown of professional associations in industrial education from the national to local level.
4. Select a particular technical area that you are interested in and make a listing of the ways in which you can become best prepared for that specialization. Make use of resource persons and reference materials.
5. Make an outline of the educational and/or occupational requirements necessary to prepare you to teach your selected technical field according to your state certification requirements.
6. Prepare a diagram indicating the organizational pattern and relationship of technical teaching areas under the main title of Industrial Education.

SPARE PARTS BINS

BOLTS

NUTS

SCREWS

GLASS EYES FOR GUYS WHO WONT WEAR GOGGLES

Chapter 2
COURSE PLANNING FOR INSTRUCTIONAL CONTENT

The planning of content for technical courses is not an easy task. It involves many factors that must be considered and evaluated as you prepare your teaching material. Over the years a variety of approaches to instructional content selection have been used to a greater or lesser degree of success. It is not here that a study of historical methods of content selection will be made, rather, a practical contemporary approach to be suggested. However, a knowledge of some of the shortcomings in course planning that have taken place in the past should prove helpful. Some courses have been planned on the following basis. It is easy to see the fallacy of each.

a. Select a textbook and follow it from cover to cover.
b. Collect a sequence of project drawings which provides the course content.
c. Use notes and activities from another instructor as course material.
d. Select activities and teach those things you like best.
e. Let the students set the pattern of the course by choosing their own activities.

As trite as it may seem, many courses are planned on one or more of the above with little teacher involvement. Perhaps too much emphasis has been placed on technical skills and teaching methodology. Although these are equally important, the evolving curriculum plan is the gateway to valuable technical courses. A negative approach has been illustrated only to point out how misleading certain plans can be. Now let's take a positive look at the many factors involved in modern curriculum planning with you, the teacher, at the center of the creative approach.

Your Responsibilities for Content Selection

To begin an analysis of curriculum planning, the term "creative approach" will be used to denote an ongoing process of content selection. The content of technical courses is not a "static" common denominator, but rather a creative design that

is open-ended, always ready for reevaluation and examination. Therefore, the development of the creatively designed curriculum and its continuous evaluation is placed directly in your hands; it is your responsibility. One of the greatest joys in teaching is revealed through the growth you see in students as they engage in and profit from learning activities that you have spent hours to prepare. Creative planning is fun, for you never

become stagnant. Better yet, you develop a burning desire to study and learn more yourself in order to keep your curriculum at its greatest potential. It is your responsibility to carefully plan course content that is in line with contemporary industrial practices. This responsibility must be taken by every industrial education teacher if new programs are to be provided. John Dewey, the famous American educational philosopher, put his finger on the pulse of today's problems of course planning when he stated many years ago:

"Since the curriculum is always getting loaded down with purely inherited traditional matter and with subjects which represent mainly the energy of some influential person or group of persons in behalf of something dear to them, it requires constant inspection, criticism, and revision to make sure it is accomplishing its purpose. Then there is always the probability that it represents the values of adults rather than those of children and youth, or those of pupils a generation ago rather than those of the present day. Hence a further need for a critical outlook and survey."

It is upon this basis that a creative design to course planning is presented. By using a creative design, the following factors should be considered while developing the total course plan:

ESTABLISH AIMS OR GOALS FOR THE COURSE. Every course plan should begin with established goals to provide a framework within which you may operate. These goals should give you direction and keep you on the "right track" as you plan your course content. Course goals will certainly vary as to the purposes of your program, whether they deal with industrial arts or industrial-technical education. Goals should be stated directly in terms of behavior changes desired on the part of the student. A few examples are listed below:

To develop an awareness of the contemporary products of the metals industry.

To develop an understanding of the installation and maintenance of metalworking machinery.

To develop a high degree of skill in the machining of metals.

Course goals, such as these, would be complimented by those more general objectives as suggested by organizations like the American Vocational Association and the American Industrial Arts Association. These would include such general topics as consumer knowledge, occupational opportunities, health and safety factors, and an understanding of contemporary American industry. You should especially make it a point to include the needs of students, modern industrial practices, and the exact purpose of your course in your stated goals.

MAKE A SURVEY OF THE TOTAL TECHNICAL FIELD ESTABLISHED BY YOUR COURSE GOALS. In order to stay with the creative design approach it is necessary to closely examine every aspect of that phase of industry your course will cover. Write out a topical diagram which gives an overall view of the content field. Now you have two points of reference from which to plan your course, your established goals and your topical outline. Whether your course is a broad industrial field such as the forest products industry, or a narrower technical area such as welding, your topical outline is a necessity. In the first example, (forest products industry), it gives you a total overview of that industry indicating where emphasis should be placed during planning. In the second example, a welding course, it provides the relationship of that technical area to the total metals field. These are extremely valuable guides. An example of a

topical outline, as illustrated for the forest products industry, brings into perspective the total components making up that industry.

Forest Products Industry

Lumbering — Saw Mill Operation — Pulp Mills

Wood Technology

Identification — Research — Structure — Utilization
Physical, Mechanical, Chemical, Variable Qualities

Seasoning & Preservation

Drying, Moisture Content, Seasoning Defects, Deterioration, Preservation

Primary Wood Industries

| Lumber and Speciality Mills | Veneer and Plywood | Timbers | Wood Chemistry Industry | Composition Products | Misc. Products Industry |

Secondary Wood Industries

Construction Industries   Wood Fuel Industries
Furniture Industries   Wood Fixtures Industries
Paper Products Industries   Patternmaking Industries
Chemical Products Industries   Misc. Products Industries
                    Transportation Industries

It is helpful to make up a topical diagram or outline sheet for any industry when preparing a course dealing with only one phase of that industry. For example, if you are planning a course in silk screen printing, you can better judge the relationship, concepts, and emphasis necessary by first preparing a topical diagram of the graphic arts industry with which it deals. Then extract from the topical diagram those phases of silk screen printing that grow out of the overall industry... "The creative design approach."

The example below, using three sections from a forest products course, shows a creative design approach being used to break down a topical diagram into teaching components:

Wood Technology

I. Wood Structure and Analysis
  A. The nature of wood
  B. Gross structural features of wood
    1. The planes for study of structural features
    2. Hardwoods and softwoods
    3. Wood rays, vessels, cells (parenchyma and tracheids), and resin canals

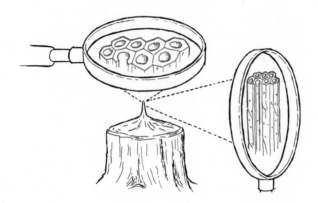

  C. Analysis of features for identification
    1. Key for identification
    2. Physical properties for identification
  D. Variable quality of wood species
  E. Natural defects in wood
  F. Figure in wood
  G. Physical and mechanical properties of wood
    1. Electrical properties of wood
    2. Thermal conductivity of wood
    3. Strength values
    4. Moisture content and strength

Seasoning and Preservation

II. Seasoning and Preservation
  A. Control of moisture content and shrinkage
  B. Determination of moisture content
  C. Seasoning of lumber - air and kiln drying
    1. Fiber saturation point
    2. Equilibrium moisture content
    3. Shrinkage factors

4. Kiln operation
5. Recommended moisture contents
6. Casehardening - its control
7. Porosity of wood
8. Density and specific gravity
9. Seasoning defects
10. Electronic drying - other new processes
D. Preservation techniques
   1. Types of preservatives
   2. Methods of impregnation
   3. Application of treated material

Primary Wood Industries

III. Primary Wood Industries
  A. Veneer and Plywood
    1. Manufacture of veneer
    2. Selection of logs - cutting of flitches
    3. Production of veneer
      a. Slicing
        (1) Horizontal
        (2) Vertical
        (3) Flat cut
        (4) Quarter cut

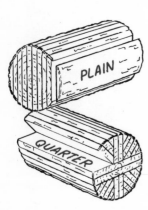

      b. Rotary cutting
        (1) Full log
        (2) Half round
        (3) Stay-log cutting
      c. Sawing
        (1) Plain sawn
        (2) Quarter sawn
    4. Grading and selection of veneers
    5. Veneer thicknesses
    6. Veneer drying
    7. The manufacture of plywood
      a. Grading and matching face plies
        (1) Book-matching
        (2) Slip-matching
        (3) Quarter-matching
      b. Taping and splicing
      c. Lumber core plywood
      d. Adhesives used in plywood
      e. Some advantages of plywood over solid wood
      f. Plywood types and thicknesses
  B. Wood laminating and bending
    1. Wood laminating in industry
      a. Structural laminates
      b. Sporting goods
      c. Boat building
      d. Furniture

2. Materials used in laminating
   a. Natural adhesives
   b. Synthetic resin adhesives
   c. Adhesive curing equipment
   d. Woods for laminating
3. Equipment
   a. Adjustable - intermittent forms
   b. Fixed - continuous forms
   c. Vacuum forming equipment
   d. Pressure forming equipment
     (1) Hydraulic presses
     (2) Pneumatic presses
     (3) Screw presses
     (4) Retaining clamps
4. Plasticizing methods
   a. Heat, water, steam, chemical
C. Lumber and Speciality Mills
  1. Classification of lumber
  2. Lumber manufacture
   a. The circular saw
   b. The band saw
   c. The gang saw
   d. Ripping and crosscutting

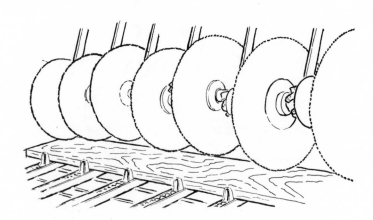

   e. Surfacing and shaping
  3. Status of the lumber industry
  4. Lumber grading
   a. Softwood Lumber grades
   b. Hardwood lumber grades
   c. Lumber measurement
   d. Grading defects
D. Wood Chemistry Industry
  1. Carbonization and destructive distillation
   a. Process for hardwood and softwood distillation
   b. Equipment

2. Destructive distillation products
   a. Charcoal - activated, briquettes
   b. Acetic acid
   c. Methanol
   d. Hardwood tar oil and pitch
   e. Acetone
   f. Creosote
   g. Turpentine
3. Wood Hydrolysis
   a. Sources of wood for hydrolysis
   b. Wood saccharification process
      (1) Dilute acid batch process
      (2) Concentrated acid process
      (3) Wood molasses production
   c. Production of alcohol and yeast
4. Cellulose derived products
   a. Cellulose filaments and yarns
   b. The viscose method - rayon
   c. Cellulose acetate
   d. Explosives
   e. Lacquer

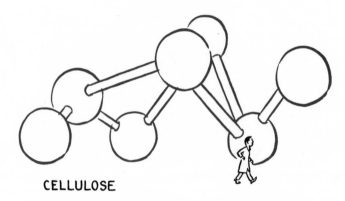

CELLULOSE

E. Composition Materials
   1. Types
      a. Chipboard
      b. Flakeboard
      c. Particle board
   2. Principal raw materials
      a. Pulp wood
      b. Industrial wood residues
      c. Manufactured material
   3. Processes
      a. Pulverized
      b. Application of adhesives
      c. Pressing operations
   4. Applications of composition board

F. Impregnated compressed materials
1. Types
    a. Compregnated
    b. Impregnated
    c. High and low pressure laminates
    d. Impregnated overlay materials

The course content outline for these three selected areas provides the necessary breakdown of subject matter to begin establishing teaching units and activities. Typical illustrations of learning activities growing out of a course content outline are discussed and illustrated in the following chapter. The term "creative design" becomes more apparent as you follow through the methods used in determining just how each individual sub-topic was selected, by you, the person responsible for content selection.

Sources of Information

Keeping in mind the sequence of developing a topical outline and your creative design for course content, the first question that should arise is, "Where would I get the necessary information?" It is hardly possible to list here all of the sources of information for content for all courses to be taught. However, for purposes of illustration, take a look at the metals industry for information for planning a topical outline and a creative course design. To review for a moment, the purpose for planning these outlines is two-fold. First, to select content that is up to date and second, to locate the relevant importance of topics within a particular industrial field and to see that no areas are overlooked.

Now, back to the metals industry. If you follow a sequence of the sources of metals, their extraction, properties, forming, fabrication, finishing and decorating, it will give you a basis of where to look for information to make your plans. The following sources of information are but an example of where to look for what you want. By not exploring as many avenues as possible, it is easy for you to overlook or not even be aware of certain things going on in the industry. It should also be definitely emphasized that most information secured and studied does not automatically become topics of content within your plan. This material must be carefully evaluated and interpreted for the level of the course you are developing. You will find that much of the information and material you secure will come from industry. Examine the following possible sources for the metals industry.

1. YOUR LIBRARY - The card catalog will list a variety of volumes dealing with various phases of the metals industry. Select and study those that provide the type of information you need.

2. INDUSTRIAL PUBLICATIONS - Many of the manufacturing and service companies in the metals industry provide publications to keep you informed on the latest developments of the industry. The United States Steel Company, Ford Motor Company and the Hobart Welding Company are typical examples.

3. INDUSTRIAL SPECIFICATIONS - Almost every company in the metals industry will supply specifications and descriptions of the products, equipment, or supplies they manufacture. A survey of those materials should be part of your educational kit for course planning.

4. ASSOCIATION PUBLICATIONS - The many professional associations within the metals industry are ready to serve you with the latest technical literature. Examples of such associations are The American Society of Mechanical Engineers, The American Welding Society, The American Society for Metals, and The American Iron and Steel Institute.

5. EQUIPMENT AND SUPPLY CATALOGS - If you were planning any metals course, you should obtain catalogs from manufacturers and suppliers which provide specifications and information on both industrial and laboratory equipment.

6. EDUCATIONAL INSTITUTIONS - Visits or correspondence with personnel from technical institutes, vocational schools, and universities can provide you with profitable information concerning the content of many phases of the metals industries.

7. TEXTBOOKS - You can obtain a good overall view of the contemporary metals industry by reviewing textbooks. At this point you are not selecting a textbook for a course, remember, you are searching for information. You should review textbooks representing all phases of the metals industry, from specific texts on foundry and sheet metal practices to broader texts on manufacturing, metallurgy, and metals science.

8. INDUSTRIAL CONTACTS - American industry is obliging. A letter or a telephone call may well put you in contact with persons in industry who can be very helpful in sharing information with you. The competent industrial education teacher usually has many industrial contacts whom he may call upon when he needs ideas or specific information.

9. INDUSTRIAL VISITS - One of the best ways you can gain an understanding of the metals industry is by visitation. Many industries will open their doors to you to view and discuss their latest operations and processes.

10. YOUR TECHNICAL PREPARATION - One of the best sources of material for course planning is that obtained during technical courses. You should constantly be aware of and collect as much information as you can concerning the content of the course to provide an "educational tool kit" for future use. Your notebooks from metals courses should be retained.

This brief listing of possible sources of information for planning course content for the metals industry should serve as an illustrative guide for you for any technical area.

You should now be able to draw together materials needed to design your "topical outline" and your "creative plan" for any given course.

Integrating Curricular Areas

Often overlooked in the development of the course plan is the necessity for integrating various phases of many curricular areas. This is not only related to technical content but also to other parts of the total educational curriculum. Very few of the technical areas of industry are isolated or unrelated to other areas. Further than just being related, many areas are overlapping and have similar adaptation to many fields. Typical examples are fastening devices, welding, graphic representation, and electronic controls. You might ask, "Should I include graphic communications in a plastics technology course?" Certainly you should, because almost everything made of plastics contains some graphic message, from a part number to a colorful display on a plastic milk bottle. A further analysis indicates the necessity of bringing together concepts and principles that are common to many aspects of industry. Properties of materials, cutting principles, testing procedures, and fabricating techniques are but a few examples of concepts that apply to many technical fields.

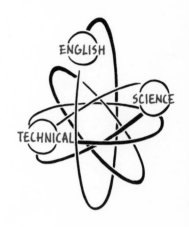

Another concept of integrating curricular areas is in the field of what might be called general education. An overemphasized but meaningful way to look at it might be through the eyes of a person who stated that, "Every high school should be named an Industrial Education High School." His intent was to indicate that every high school depends on industry or is related to industry throughout its total curriculum. Someone, of course, challenged this with the comment, "Just show us how music is related to industry." This was an easy question. If industry did not exist, there would be no instruments, no printed song sheets, no high fidelity recording equipment, no music books - - perhaps a quiet music class sitting on a dirt floor. Overemphasized, yes, but it does indicate how the threads of industry are woven throughout the total school curriculum. Perhaps a short list of illustrations will help you see the relationships:

    1. There is probably more English used in technical writing and communicating than in any other field.

    2. Science is the foundation upon which industry is built.

    3. More manipulative skills are required in technical areas than all other fields combined.

    4. Mathematics and measurement are "tools" in the industrial world.

    5. The social sciences revolve around the occupational pursuits of our society.

It is your responsibility to incorporate the concepts and principles of all related educational areas of study into your course plan. Most of these will be revealed as you interpret the many sources of information while preparing the content for your course plan. Your course is not an isolated segment of knowledge.

You and Textbooks

Regardless of the type of course you are planning to teach it is essential that each student has a textbook available. The textbook may or may not belong to the student, but in either case it is helpful if he is able to use it at home for outside study and assignments. You will notice that the selection of a textbook has come after the development of the course plan. It has already been indicated that the use of many textbooks is valuable as a part of your source of information for your planning. However, now that your plan is completed it is necessary to select a textbook that compliments the content and provides the student with concepts, facts, and principles as closely aligned with your plan as possible. Three main factors aid in textbook selection:

a. The reading and concept level should be geared to the level of your class.
b. Basic content should lead outward to involve further study and investigation along the lines of your course plan.
c. The subject matter and organization should be contemporary and representative of modern industry.

Even with all the effort you may make toward evaluation of textbooks for a particular course, it is doubtful that you will find exactly what you feel you need for your students. This is to be expected if you have followed a type of course planning as has been suggested. No textbook was written specifically for your course, so your choice becomes that of a selection which comes the closest to meeting the goals and outline of your course.

You and the textbook are the human and static resources surrounding the student as he engages in the activities and study provided by your course plan. Do not let the textbook be the course, but rather, use it like any other resource material as discussed in the section on teaching activities.

Contemporary Analysis for Content Selection

It is essential that a creative plan for course content, evolving from a topical outline, is formulated through a contemporary analysis of industry. The outlines and plans that have been illustrated previously require this type of approach. This is not new, but perhaps the action and sources are new. Too often a vocational or industrial arts teacher may be teaching what he "thinks" is being done in a certain industrial situation rather than what is actually being done now - today - or better yet, tomorrow. A contemporary analysis is based on today's industry, its research, and extensions into the future. As a teacher planning any kind of a technical course, you should develop your course plan in such a way that every student has the opportunity to learn in the context of the latest happenings in that phase of industry. You must realize that new materials, new techniques, new processes, new ideas, and new designs are pouring out of industry every day. In fact, it is a demanding task to keep up with industrial innovations, but the searching into these innovations and reorganizing them into content for student study is the contemporary analysis. It requires a continual reevaluation and examination of your course plans according to the new information and material you receive through the various sources previously discussed.

Your Relationships with Industry

Probably no other factor can be emphasized more in contemporary course planning than your relationship with industry. This is the prime means by which you grow and live with your technical area of teaching. It is a good idea to establish lines of communication with industry to help you get information and solve problems. These lines of communication should come in many forms. Although it takes time, the established teacher will have a variety of contacts at his fingertips. To illustrate this point, a few examples follow:

a. BE A LETTER WRITER - before long you will have many contacts in industry, people who are always willing to assist you by providing information, materials, and even equipment. These people know their business and are usually happy to share their "know-how" with you.

b. BE A VISITOR - every opportunity you have to visit an industry in your technical pursuit, do so. A personal contact is

often more valuable to you than a written contact, for these persons may better understand your program and your needs as you visit with them. Industry is interested in education, for your program may be a source of supply of educated employees.

c. BE A SUBSCRIBER - get your name on the mailing lists of manufacturers and suppliers in your technical area. This is the means by which industry communicates within itself, shows its products and services, and you can become part of that system. Also, subscribe to periodicals in your field. As each issue is received, it is amazing how much new material you gain as you read the articles and advertisements.

Your relationship with industry provides one of the major avenues through which you can gain a contemporary analysis of the ongoing changes and developments in your technical field. Keep this avenue open for it provides many opportunities for you. As well as assisting you in course planning through a knowledge of new processes and information, this relationship also gives you an insight into obsolete material that you may want to delete from your course plan. A program so inspired can never be classified as out-of-date. You can establish a basis for course planning, and then as your teaching career continues your relationship with industry becomes your continuing in-service education program.

Resourcefulness - Your Key to Good Planning

If there were one characteristic that could be identified as the key to the best in course planning it would certainly be RESOURCEFULNESS. The resourceful teacher uses every means at his command to come up with the ultimate in a totally planned course. He uses all of the contemporary techniques previously suggested, and many more, to be sure he has not overlooked any aspect of planning. You may give thought to this reaction of an administrator in industrial education when asked about employing a new teacher in electricity-electronics. "Find us a person with some background in this technical area... one that is resourceful." When questioned further, he indicated that he believed a resourceful teacher had no traditional restrictions that would limit or hamper his approach to contemporary course planning. He would apply himself, add to his technical background as he went along, and develop outstanding courses for the students. Whereas a traditional teacher in electronics may not be able to overcome his view of the field and do little to continually up-date his courses.

There is much sense to this viewpoint. Contemporary planning for course content relies heavily on the resourcefulness of the teacher. In short, the resourceful teacher does not have all the answers. He does, however, have the ability and desire for searching and evaluating all aspects of possible content for creative courses. You should constantly look to every resource available in your quest for content selection. Plan your courses, not just around what you know, but around what you can find out by being resourceful - - your key to good planning.

Revision is a Must

Contemporary analysis for course planning is not enough in itself. It must be followed by constant evaluation and revision. There are two aspects to revision. First, the course plan is actually tested by use. There may well be many aspects of the course that looked good during the planning stage but did not work out well during the actual teaching. Awareness of any inadequacies in course content as the course is being taught should be noted for revision for the next time around. This type of revision requires you to evaluate the content after each class meeting, making notes and suggestions for future changes. If you wait until the end of the course you may not remember all of the details you would like to have on hand when planning the next session.

Second, revision of your course plan should be an ongoing process in regard to technical advancements. These revisions should come about in the same manner as your original course planning. You must constantly be aware of innovations in your technical speciality which you can interpret and incorporate into your plan. Just as you expect your students to follow assignments and study the course activities, they have the right to expect that you have prepared a practical contemporary course for them. Your courses will never be out-of-date if you keep in mind that continuous revision is a must.

Units for Lesson Planning

The next logical step in course planning is to divide the content into related teachable units. In this way you give the student an opportunity to see the course as a whole through the interconnected units of your plan. You also put the course into perspective for the student as he progresses and relates one unit to the next. When teaching, you should provide the student with a copy of the units making up the course so he is able to see the total plan of study.

The following unit outlines for specific courses will provide you with examples of how a course may be divided into sequential units.

Practical Electronics

Unit 1   Introduction to Understanding Electronics
Unit 2   Electric Energy Sources
Unit 3   The Flow of Electrons
Unit 4   Tools and Materials in Electronics
Unit 5   Magnetism
Unit 6   Electrical Tests and Measurement
Unit 7   Systems Used in Communications
Unit 8   Radar, Radio, and Television
Unit 9   Transistor Electronics
Unit 10 Electronic Manufacturing Industries
Unit 11 Careers in Electronics

Plastics Technology

Unit 1   Introduction to Plastics
Unit 2   The Plastics Industry
Unit 3   Basic Chemicals for Plastics
Unit 4   Manufacture of Resins
Unit 5   Properties of Plastics
Unit 6   Compounding and Available Forms
Unit 7   Molding Processes
Unit 8   Mold Design and Construction
Unit 9   Fabricating and Decorating
Unit 10 Factors in Material Selection

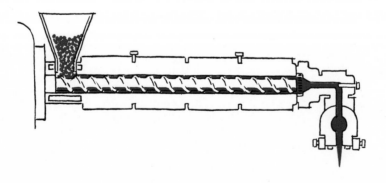

Each of the units in a course refers back to your total course plan. Detailed subtopics may be used to expand each of the units into your teaching content. The units of a course may or may not be the same as the headings in your topical outline. The topical outline provides you with information from which you should organize your units. A unit breakdown of a course plan is strictly an individual affair. It depends mainly upon the organizational patterns that each individual teacher prefers. In other words, the same course plan may well be set up in different units by different teachers with the same end results.

Now you have one major portion of your teaching plans ready for the preparation of lesson plans. Each unit in your course represents a great deal of content which requires the development of a number of teaching-learning activities to bring out concepts, develop skills, and provide an understanding of principles. As the development of learning activities is discussed, you will have an opportunity to relate the concepts of learning activities and content in providing the best student learning situations.

Organized Contemporary Programs

In the past attempts have been made by individuals and groups to establish completely organized programs for various school levels that cover broad aspects of industry. Some of these programs had little influence on industrial education while others have had a considerable impact. Undoubtedly the influence of such programs has been dependent on such factors as funds for promotion and publication of materials, organized teacher preparation centers, and supervision of such programs in the schools.

In recent years a number of projects have been developed which have had the support of educational institutions, government agencies, and school systems to the extent that they are having a profound impact on industrial education programs. You should become familiar with the major projects, first to see if they may fit into your teaching situation and, secondly to become knowledgable of all contemporary industrial education. None of the contemporary projects can be expected to replace industrial education, but rather to compliment existing programs and give the student a better opportunity to gain another dimension of learning and industry. A brief outline of some of the major programs follows:

The Maryland Plan

This curriculum plan is based on an anthropological approach to the study of basic elements common to all civilized mankind. It includes the study of power and energy, tools and machines, and communication and transportation at the beginning level. It involves the general study of industry as it relates to man in his industrial environment.

The second level is geared to a contemporary approach to the study of American industry through group learning processes. Activities involve in-depth study of a selected industry using the group project approach and similar study of industry using the line production approach.

The third level of the Maryland Plan involves personal emphasis on the psychological needs of the individual and developing his resourcefulness. Special emphasis is placed on the development of mental problem solving skills through the medium of research and development. The outstanding aspects of the program are in the interrelationships of man learning how to solve problems in a contemporary industrial setting.

You should take an opportunity to investigate the details of this project from the reference section, page 193. The aims of a typical program based on this plan are as follows:

1. To acquaint the student with the world of materials through media presented in classtime activities.
2. To develop a basic understanding of the material processes as are used in industrial situations.
3. To enable the student to become familiar with the structure of materials as applied to its uses.
4. To encourage the student to learn content pertinent to modern industrial products and practices.
5. To familiarize the student with the history and development of materials in both the educational and industrial fields.
6. To create an understanding of research techniques used in procedures and processes in the industrial world.
7. To provide an opportunity for independent study in research and experimentation activities.
8. To emphasize problem-solving situations in areas of content experiences and procedures as may be found in the world of materials.
9. To transfer learned knowledge as needed to solve problems in practical situations.
10. To assist the student in developing an awareness of design to meet a specified need.
11. To advance responsible and serious investigation of the future through class activities.
12. To facilitate communication and cooperation among individuals while completing class assignments.
13. To promote the development of methods for the study of present day and future technology through research in industry and education.

American Industry Program

The industrial education program known as American Industry is an outgrowth of a proposal made at Stout State University to provide students with an understanding of industry not normally found in other technical programs. It has been supported by federal and foundation funds, and through the efforts of a team of educators, has been formulated as a four year degree program in teacher education. Follow-up and supervisory personnel have contributed to the continuation and refinement of the American Industry program.

The basis of the program lies in a conceptual structure of the knowledges of industry. By dividing industry into concepts such as management, research, finance, processes, marketing, and so on, a course outline is planned that will closely resemble the functional operations of a typical American industry. In the laboratory, the students assume responsibility for the organization of an industry and perform all the necessary functions associated with the production and marketing of a product or service.

This type of industrial learning situation provides the student with the opportunity to gain experiences in the total industrial environment. The use of tools and equipment is not minimized but becomes a means for solving problems, from management and product design, to market analysis and distribution. Every student has the opportunity to gain conceptual understandings of the organization of industry from meaningful learning activities. Details of the American Industry program may be found in the reference section, page 193.

Industrial Arts Curriculum Project

The Industrial Arts Curriculum Project was proposed by a group of educators at Ohio State University to develop an educational program for today's youth based on modern industrial technology.

The program consists of a two year sequence experience in the World of Construction and the World of Manufacturing. Collectively, these two areas of study provide the student with understandings and concepts related to various aspects of the wide world of industry as the dominant factor of society. The program includes student textbooks and teachers guides for each area. The textbooks contain reading assignments, laboratory activities, research, and evaluative devices. It is a highly structured program that covers various aspects of the industries, yet it is flexible enough to provide for individual differences, group and individual activities, and self evaluation. Information concerning teacher preparation, instructional materials, and teaching centers can be found in the reference section, page 193.

Review Questions - Chapter 2

1. What are the major responsibilities of the industrial education teacher regarding content selection?
2. Why should planning for any technical course begin with the development of a topical diagram?
3. What is meant by the "creative design" approach to course planning?
4. What factors should the industrial education teacher try to avoid when beginning his course planning?
5. Why is it important to consider the integration of curricular areas when planning a technical course?
6. What factors should be considered when selecting a textbook for any given industrial education course?
7. Why turn to industry for information as the main source for content for technical courses?
8. What does a contemporary analysis for content selection demand?
9. What is the major key to good course planning? Why?
10. What factors contribute to the development of teaching units out of your course plan?
11. What are the two major aspects of course plan revision? Why are these important?
12. Why is it important to keep lines of communication with industry open?
13. In what ways can you establish good lines of communication with industry?
14. What is a contemporary analysis of today's industry based upon?

Suggested Activities

1. From your technical area of specialization or interest, make a survey and list of industrial publications that would aid in course planning. Be careful to use all sources of information to locate such publications.

2. Design and chart a topical diagram for the industry in which your technical specialization lies. Talk with your instructors, use reference materials from your library, and where practical, visit industries to secure information.

3. Prepare a number of examples that illustrate the involvement of industry in any area of academic or general education other than technical courses. Indicate how these areas or courses would be affected without the influence of industry.

4. Try to make a case and give support to the concept that, "General education is a part of industrial education." Plan your paper as a reaction to the older concept that industrial education is a part of general education.

5. Prepare a listing titled, "sources of information," for your particular technical interest. Review all possible sources that would aid in the development of a course planning activity for a selected technical course.

6. Make a list of contemporary textbooks from which a selection could be made for a course you may be planning to teach.

7. Take one section of your topical diagram and develop it into subtopics which would provide for teaching content.

8. Develop a set of aims or goals for a selected course in which you are particularly interested. Use a wide source of information and references in planning these goals.

Chapter 3
DEVELOPING
LEARNING ACTIVITIES

Implementing your course outline so maximum learning results, is undoubtedly the most challenging part of teaching a technical course. The implementation of the course outline relies basically upon your planning of learning activities. The development of learning activities is similar to course planning in the sense of using a contemporary approach. Modern learning activities that provide the greatest opportunity for student understanding and comprehension should be your goal.

The psychology of learning has slowly shifted in emphasis over the years from teacher domination to greater student involvement. In other words, the responsibility for learning has been placed more in the student's hands, while the teacher has become more involved in planning learning activities that provide for motivation, opportunity, and responsibility. This goes along with the common educational theory that you really do not teach a student, but rather, you provide an opportunity for him to learn. It is within this framework, then, that you should approach the task of formulating the variety of learning activities necessary to bring out the desired concepts, skills, and principles of your course. This requires you to carefully examine the details of the content from your course plan and make decisions as to what type of activities will help the student best achieve his goals and the total goals of the course.

CONCEPTS IDEAS

CHOICES

PLANNED ACTIVITY

Learning activities may be classified as covering a wide range of teacher-student involvement. You will find there is no one type of learning activity that will solve all learning situations. Rather, an analysis of each concept or skill will narrow your choice of activities. For example, you may want to teach the concept of electrical resistance in wire. Now you must design a learning activity that will best bring this concept about in the mind of the learner in the least amount of time. You have many choices, all of which will not be satisfactory. Your job is to devise the best activity possible. Briefly, look at two extreme activities for our electrical resistance concept. First, an assignment to make a soldering iron. Secondly, an experiment to measure the resistance of selected wires of various gauges and materials. These are extremes but they make the point. The soldering iron would take hours of manipulative work, little of which would deal with resistance even if it were stressed. The measuring experiment would deal directly with the desired concept to be learned and save a great deal of time for other activities. The point is not that a soldering iron makes a bad learning activity. It is just that it did not satisfactorily meet the needs of the stated concept to be learned. An analogy might be learning how to fly an airplane to understand the principle of providing power from heat developed in an engine.

It boils down to the fact that you have quite a responsibility to plan learning activities that meet the needs of the situation and eliminate "busy work." There must be some excellent activity for any skill or concept to be learned. Your job is to locate it and refine it. In the following paragraphs we will discuss a number of selected learning activities which illustrate how content and learning may be coordinated.

Selecting Activities for Meaningful Learning

In general, all learning activities have a number of features in common. The term "activity" refers to action on the part of the student as he engages in the learning process - - not a lecture, demonstration, discussion, or reading assignment which indicates passive learning for the student.

PURPOSE

EQUIPMENT

MATERIALS

PROCEDURE

ACTION

EVALUATION

Some of the major features of well planned learning activities are as follows:

State the purpose of the activity clearly and specifically. Formulate a title that is meaningful. List the necessary pieces of equipment and materials for the activity and have them ready for use. The procedure or presentation part of the activity should indicate the sequence to be used and all necessary information so the student may proceed on his own. Your activity should also include some method of evaluation, such as discussion questions, recording of data collected, check points in construction, or comparison testing.

With these factors in mind you can begin a selection process for planning learning activities to meet the goals of your course plan. It is here that you go over your course plan, unit by unit, and make the decision as to the type of laboratory activity that will provide the most meaningful learning situation possible for each topic.

Selection of learning activities should also be established on a "psycho-logical" basis for sequential learning. By "psychological," reference is made to learning situations that are logical to the mind. The student sees the immediate necessity and order of what he is to learn. He can also make immediate use of the mental and manipulative skills he has learned. The student's background, interests, and needs are provided for when you plan your learning activities on a psychological basis. The activities will also lead him into wider realms of experience and interest.

Too often the activities for learning are established in a "logical" sequence which does not take the student into account. In a logical sequence the content is arranged in building blocks from easy to difficult, or from simple to complex. Often the student does not comprehend what he is expected to learn nor is he able to make immediate use of it. It is advantageous in that it is undoubtedly easier to plan, however, it is less likely to arouse and hold student interest. Typical examples of the logical sequence have often been noticed in technical teaching. The concept that a student must learn woodworking before metalworking because it is not as precise or difficult is an illustration. Another may be requiring the student to become skilled in using a handsaw before allowing him to use the circular saw. Does one really lead to the other? Students often become uninterested when they do not see the need or use of such logically arranged content.

Manipulative Activities

Learning activities in which the student is engaged in creating, doing, or manipulating, take advantage of many of his senses. The more senses involved in a learning situation, the more opportunity and probability that understanding will result. Learning by doing may be trite, yet no other term more accurately describes the activity part of the learning process for it adds the senses of touch, smell, and taste to the overemphasized senses of seeing and hearing. The industrial education laboratory, with its great potential for manipulative learning situations, should be carefully exploited.

Although there are some who advocate a certain type of manipulative activity to reinforce course content, it would be well for you to examine the many kinds of activities which will help the student grasp all aspects of the subject matter. Rather than the construction of many products to fulfill the necessities of laboratory work, you should be aware of the broad spectrum of manipulative activities that are possible to make your teaching the most productive in all aspects of learning. You should give your attention to activities you can plan from completely teacher oriented all the way to completely student oriented situations. There are two main reasons for using a wide variety of activities. First, all laboratory activities of one type do not lead to the development of the mind to the many ways of approaching and solving problems. Second, it is doubtful if you will be able to interpret all of the course content through one media.

Now, let's take a run through a typical industrial education laboratory and try to visualize, through a number of illustrated activities, how you may begin planning. Remember, the concept you are attempting to present should be represented by the best activity you are able to devise. There are many to which you can attach names, such as research activities, tests, product construction, experiments, evaluations, comparisons, exercises, and equipment operating instructions. Others may be more elusive of a name, or a combination of those mentioned. Let's look at a number of examples which should prove useful in your activity planning:

A Problem Activity

The concept of planning student problem activities gives you an opportunity to get across certain important topics of your content as well as evaluate what the student has learned. You also can gain an understanding of how students approach and solve a given problem. An analysis of the problem solving method of teaching is discussed in detail in Chapter 8. This should be helpful to you in organizing and presenting this type of activity.

In order to keep your students on the right track and avoid misunderstandings, it is wise to prepare a problem activity sheet. The student may then refer to this material at any time he is in doubt or has questions concerning any aspects of the problem. With the problem in his hands and mind he is ready to act. If you simply state the problem and tell him to "get busy," you may find he is unable to remember all the details. Give your students the necessary information in written form and save yourself more time to teach.

Plan your problem activity sheets so the content is on the student's level, new, and challenging. A well planned problem activity can provide a great deal of interest and motivation. Examples of problems conducive to this type of activity might be as follows:

1. Design a device to hold a set of kitchen carving knives.
2. Make a layout and describe methods to be used in printing some personal calling cards.
3. Find the average weight of a selected group of materials.
4. Make a chart to convert linear measurement up to one foot from the English to the metric system.

These few examples should give you some ideas of the nature of problem activities. Any technical area is full of content which lends itself to problem activity learning. Your problem is to identify such activities from your content and reorganize them into a good teaching-learning format. A problem activity sheet should include the following:

1. A title for the problem.
2. A general description of the problem.
3. Directions for the use of necessary materials and equipment.
4. Instructions concerning completion of the problem and how it will be evaluated.

The problem activity is teacher orientated and sets the stage for student thinking, planning, and action. Select your student problem activities with care, make them meaningful to the student, and relate them directly to the goals of your course. In other words, let the student know why the problem is so important and what he will gain by engaging in the activity. Tailor the problems to the capacity, needs and interests of your students.

Product Construction Activities

The major emphasis of product construction activities is based on "doing" to help the student better understand what he is learning. Just to make something for the sake of keeping busy, having fun, or because mother would like it, is out of line for your planning of construction activities. These activities should be derived from your course content and planned by you for three prime purposes:

1. To reinforce abstract learning by practical application.
2. To develop manipulative skills in using materials, tools, and equipment.
3. To understand the relationship between planning and doing.

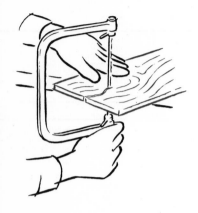

Constructing products that meet these requirements is just one of the many vehicles you have at your command to assist the student in better performance and understanding. The vehicles should be directly related to, and part of, the flow of sequential learning of your technical content. You do not fill one part of the student's brain with information and another part with skills. The best learning takes place when all parts of the learning system are activated and tied together by good teaching. Therefore, your selection of products to be constructed requires an analysis of their value to the overall learning objectives. Construction activities are essential. Make appropriate use of them.

You may require students to construct a corner shelf, a set of laminated salad servers, a model home, a chess set, or a letter opener. It's not so much "what" the student makes but "why" he is constructing it and, more important, what skills he will develop and what he will better understand. Then product construction has real meaning for learning. Construction products selected on these factors will keep your students "busy," they should have "fun," and mother may very well enjoy the results.

Construction of projects has undoubtedly been over emphasized in industrial education courses with disregard for many other desirable learning activities. Complete domination of one type of learning activity to bring about student performance and understanding usually does not provide the student every opportunity to learn. It is somewhat like teaching a child to ride a bicycle by reading books and articles. That is part of the process, but other activities such as experimentation and performance will also aid in completing the picture of a good bicycle rider. Give your students the same opportunity. Put product construction in the correct perspective as one of many parts of the learning system. As you do this, be sure it is being used to bring about the desired learning behavior for which it is intended. For example, your content may indicate a need for student understanding of wood identification, so you plan an activity for each student to build a coffee table. Get the idea? There must be a better planned activity for learning wood identification. The coffee table may also be a fine activity for some other phase of content. Like a game of mix and match, be sure you come up with the right activity for specified learning concepts.

Product construction presumes a need for manipulative skill development along with an understanding of principles and concepts. It is also assumed that the student has obtained the necessary background to successfully complete the product and yet be challenged by its accomplishment. The following features should be carefully considered when planning a product construction activity:

1. Provide an accurate name or title for the product to be constructed.

2. Give a statement of the purpose of the construction activity. Make it precise and explicit as to the objective for performing the activity.

3. Indicate the necessary materials to be used for construction of the product. Provide accurate specifications.

4. Specifying a plan of procedure. Guide your students. Do not leave them wondering what to do next.

5. Provide the necessary drawings of the product. Your drawings should be neat and accurate to convey the correct ideas to the students.

6. Indicate a section for student evaluation of his progress and difficulties he experienced during construction.

It is also suggested that an activity sheet be prepared to provide the student with the necessary procedures and instructions.

Activities for Operating Instructions

Many pieces of equipment or machines used in technical courses are complicated and require the student to check for specific operating instructions. In such cases it is extremely valuable to devise an activity sheet to aid the student in using the correct operating procedures. Often you can plan this type of activity by using the manufacturer's specifications. It is especially important that sequence of procedure and safety precautions be emphasized.

This type of activity planning assists both you and the student in a number of ways. It provides guide lines for students to follow in learning the importance of following a sequence in the operation of some equipment. It also points out to him the concepts to be learned about certain equipment which he may not remember from a demonstration. Another advantage is that damage to equipment may be minimized. Too often equipment is broken because a student did not follow proper instructions. Your teaching may also be more effective. If students are following operating instructions, you can do more teaching rather than supervision. In some instances teachers spend a great deal of time overseeing student equipment operation so that equipment is not damaged and students are not injured by following improper sequences.

It is your responsibility to make decisions as to when activity sheets for equipment operation are necessary, both for student learning and safety. These decisions are usually based on how complicated the equipment is. For example, it is doubtful that such sheets would be necessary for using a handsaw or a portable power drill. However, it may be just the ticket for a small bench model plastics injection molding machine. The student may rely on his activity sheet for information on how to load the machine, make correct temperature settings, insert molds, plan pressure requirements, and many others along with safety precautions. You make the decisions. Activity sheets take time to prepare but they may result in better learning, greater safety, and more efficient use of your time as a teacher. Try the following suggestions in preparing activity sheets for operating instructions:

1. Provide a title for the machine or equipment.
2. Indicate any necessary setup or beginning adjustments.
3. Outline any materials to be prepared or secured.
4. Present a specific procedure to be followed.
5. Insert any safety features as they appear appropriate.
6. Indicate how the equipment should be shut down and readied for future use.

Machine Maintenance Activity

Desired learning concepts are often provided by laboratory activities which involve the student in machine maintenance and repair. Your content outline and type of course will help determine if such activities are appropriate. In this respect, your responsibility is to review your course content as you are planning your teaching material and identify those items which deserve consideration for student activity. Any maintenance or repair in which students are to be involved should be directly related to the objectives of your course. They should involve concepts or performance which bring about better understanding and learning in line with the technical material being studied. A few examples may better serve to illustrate the value of maintenance and repair activities; those you should do yourself and those which may be appropriate for your students.

Preplanned activities become part of your teaching outline. In a woods course you might plan for students to weld a broken band saw blade or make new jaws for bench vises. In a metals course, your students might grind twist drills, repair lathe centers, or heat treat a screwdriver blade. A course in electricity is full of content for the maintenance and repair of equipment. Is the point clear? When the activity lies within the scope of your course, the learning outcomes may be very beneficial for the student.

Be careful not to stretch the point. Use good judgment and consideration. Students must not be exploited to lessen your responsibilities. An activity for students to paint tables in a drawing course is usually as unjustified as teaching automotive tune-up in graphic arts. Planning is necessary. On-the-spot maintenance may be educationally sound when you have a preplanned activity sheet for students to follow. Students are then made aware of the value and learning which may result from the activity. Suggested guide lines for the preparation of an activity sheet for students to use in maintenance are:

1. Use a title such as, Equipment Maintenance Report.
2. Provide space for the student to name or identify the machine or piece of equipment.
3. Have the student indicate the nature of his performance. It may be preventive maintenance or corrective maintenance.
4. Provide adequate space for learning experiences the student feels he has obtained from the activity.
5. Make room for a statement of your evaluation of the student's performance to help him evaluate himself.

Remember, do the job yourself if it is your responsibility. Plan for student activity when the job falls within the scope, time limitations, and objectives of your course.

Experimentation Activities

From an evaluation of your course outline, you may often find that student laboratory experimentation will bring about the desired learning outcomes better than other activities. Experimentation is one of the leading methods for student learning used in the sciences and engineering. The possibilities for its use in industrial education courses has not scratched the surface of technical content. Undoubtedly some of the traditional practices of teaching industrial education courses has hindered the use of student experimentation as a desirable learning activity. Emphasis given to the take-home project and skill development has left some voids on the part of student learning. The purpose of using a variety of activities in your course is to fill learning voids for your students. Experimentation is one of the best "fillers for learning" you have available. It makes it possible for students to attain important concepts that "making something" will not bring about. And yet, making something is usually involved in all types of student experimentation.

To establish a framework for student experimentation activities, it is important to design an activity sheet which will serve to guide the student toward the desired goal. Designing such an activity sheet presumes a definition and knowledge of the term experiment. Actually, two forms of experimentation may be used. First, a planned experiment through which the student is guided to known results. These have been performed before and results established. The purpose of this form of experimentation is to provide the student with an opportunity to gain an understanding of the concepts involved. Take heat treatment of steel as an example. An experiment may easily be designed to assist the student in bringing together seemingly unrelated facts and develop an overall understanding. "This is a concept." Your experiment might carry him through an understanding of hardening, tempering, normalizing, annealing, and stress relieving of steel, from unrelated facts to an understanding. This has been done many times and the results of each is a known fact. Yet an experiment may enable your students to understand the concept and not just the result; that the steel became hard. This may be applied to many topics of your technical content as the best way for students to learn. It is not always necessary to make a chisel or center punch in order to teach heat treatment.

The second form of experimentation for student learning follows the same procedure except that the results are usually unknown in advance. The same principle of concept learning applies. For example, an experiment activity is set up to find out how different types of paper react to a selected group of printing inks. Selection of papers and inks would be made at the time of the experiment. A student may have found some papers not even on the market yet. So the experiment progresses to a final evaluation of the results when conclusions may be made. As a teacher, you would not know the results nor would the students.

What an opportunity to learn. Try to look inside either of these examples. If you look closely, you would probably observe students performing, using all kinds of tools, materials, and equipment. They would be thinking, analyzing, comparing, resourcing, reflecting, evaluating, and concluding. It could well be compared with the same type of experimentation that eventually landed a man on the surface of the moon. Give your students the opportunity to do experimental activities. You will be giving them a real opportunity to learn. Some suggestions for planning your experimentation activities are:

1. Set the atmosphere for experimentation by discussion and questioning concerning the content being studied.
2. Prepare your activity sheets with a descriptive title.
3. Indicate the exact purpose of the experiment.
4. Make a listing of the necessary materials and equipment to be used. Specify quantities and sizes.
5. Outline a procedure approach to guide the student in his performance.
6. List references which may be of value as resource material.
7. Prepare guide lines for student evaluation.
8. State questions to assist the student in reviewing his activity and think through the concepts he has learned.
9. Allow for the recording of results and conclusions.

Testing Activities

The use of test methods in many technical courses is one of the best learning activities available to bring out important principles and understandings on the part of the student. You will find standard testing procedures for most technical areas in literature and reference materials. Many tests require highly sophisticated equipment while others use relatively simple apparatus. Equipment available may limit testing, but do not give up. Many pieces of test equipment can be constructed in the laboratory.

As with most student activities, testing is more adaptable to some technical areas than others. Again, it cannot be over-emphasized that your responsibility is to identify concepts to be learned and then select the most appropriate activity to bring them about in the mind of the learner. Testing provides the student with an opportunity to better understand material properties, physical principles, electrical measurements, and many others. In most instances, the student is again involved with manipulative skills, use of tools and equipment, and a variety of materials as he prepares test samples. This concept is similar to your use of a variety of teaching methods. They are overlapping. Although they are given different names, you should observe that they are not isolated activities, unrelated to one another. During a given testing procedure a student may also be constructing, researching, and following operating instructions.

Testing activities are somewhat similar to experiments. The major difference is that testing is used primarily to find physical, chemical, or electrical properties. For example, it may be necessary for a student to test the hardness of a piece of steel, moisture content of a board, chemical resistance of a certain plastics sample, or the strength of a glue joint. In such cases, you should provide for testing activities so your students may obtain results and make comparisons with industrial standards. Student testing activities provide exciting learning when they are integrated into the ongoing content of your course. Testing also reinforces many concepts that other activities or methods of teaching fail to provide. For example, you may want your junior high students to understand the importance of dry lumber and its relation to moisture content. You can define moisture content, discuss it, make explanations, show examples, or use any other methods. Do they really understand the concept? A simple, well-planned moisture content test, performed by the student, will probably bring about the understanding with more concrete meaning and in less time than you would expect.

The question is often brought up as to where a teacher can locate important testing procedures for his technical course. The obvious answer is; in textbooks, reference books, industrial literature, and from industrial visits. It is quite important that you select the desired testing activities and reinterpret them in the form of an activity sheet at the level of your students. Many complicated tests may be easily simplified and yet retain the desired concepts to be learned. In your preparation of testing activity sheets, consider the following format:

1. Use a title that clearly describes the test.
2. Provide an introductory statement which illustrates the importance of the test and defines the title.
3. Make a statement as to the purpose of the test in relation to your course content.
4. List the materials and apparatus to be used.
5. Outline the procedure to be used in performing the test.
6. Provide space for recording results and calculations.
7. List some discussion questions which will assist the student in reviewing the test and reinforce concepts to be learned.

Student Planned Activities

In order to provide for reflective thinking and problem solving, you may consider the use of student planning activities. The use of planning activity sheets gives the student some guide lines for his study. Activities of this nature allow the student a great deal of freedom and help him to gain a perspective of how to approach problems of particular interest to himself. You, as the teacher, must certainly act as a resource person in the planning, doing, and evaluative stages. The student, on the other hand, must be able to accept the responsibility for the development of the total activity. Considerable time is often spent in preparing the student for problem solving with activities of this nature. Too often a student "takes off" on the construction of a product or other performance activity without enough planning to successfully carry it through to completion. Such lightly planned trial-and-error learning may become frustrating to the learner, place an undue burden on the teacher, and result in ineffective learning.

Student planned activities have two valuable advantages. First, the process of planning, if done correctly, incorporates many learning experiences in itself. Second, the student has an opportunity to perform a variety of activities which may well be suited to his particular needs. Your supervision is necessary for both phases. However, a well-designed planning activity sheet will shift much of the responsibility from you to the student. The student's planning and his performance become his responsibility. Some students will be able to accept this responsibility. Others may have a difficult time. Your insights into your students' abilities and past performances will help you decide when this type of activity will be most meaningful. It is definitely a problem solving situation. A study of the problem solving method of teaching in Chapter 8 should be helpful in preparing your students for this type of individual activity. Consider the following factors when you are designing a student planning and evaluation sheet:

1. Title - Planning and Evaluation Sheet.
2. Activity or project name. Provide space for the student to state the problem.
3. List the functions. You may use questions to guide student thinking, such as: What are you trying to accomplish? What is special about it? What is it supposed to do? Or, how is it to be used?
4. Tentative solutions. Suggestions for the student to get ideas, discuss, make sketches, and do planning.
5. Proposed solution. Suggestions for a plan of attack, finished sketches, completed drawing, templates, or other media.
6. Materials required. Provide space for the student to list his material needs. The amount, description, sizes, and cost may be appropriate.
7. Evaluation. Assist the student by stating questions for his own evaluation. Examples might include: To what extent did you accomplish your goals? Was the solution used the most effective for the problem? Considering function and appearance, how could the design be improved? What subject matter, concepts, techniques, or skills did you learn from the experience?

Here you have had an opportunity to analyze a few of the many possible types of laboratory learning activities that may be planned for technical content. Remember that one of the most important aspects of manipulative activities is that they are planned to make use of the best learning techniques for the particular principle, concept, skill, or understanding to be learned. The appropriate use of such activities gives the student the opportunity to grow and develop in many ways. Think back. Do you realize how many different ways the student could use his mind and physical senses, plus manipulative coordination, if a variety of each of these activities were provided for him in a given technical course? He would have used his eyes, ears, and sense of touch. He would have done reflective thinking, planning, data collecting, research, designing, evaluating, creating, estimating, analyzing, diagnosing, constructing, experiencing--the list could go on and on.

These things should be on your mind when you are planning learning activities with the goal of "giving the student an opportunity to learn."

Group Activities

Many of the activities already discussed could be altered and used in group learning situations. However, there are many other activities that qualify nicely for group study. You can readily see the value of group learning activities in such dimensions as the social aspects, understanding of cooperation in industry, automation, team research, mass production, and data analysis. Again, some course content is more appropriate for group learning than other and the decision is yours to make, based on the goals of your course.

Group learning activities give the student an opportunity to realize how much workers at all levels must depend on each other to produce the desired results. Other educational values derived from group activities are:

1. The realization of progress through the sharing of information.
2. The rapid development of ideas through "brainstorming" and team research.
3. Techniques by which people must work together to gain maximum results.

You should take every opportunity to closely examine your course content to isolate those bodies of knowledge or understandings which can best utilize the group activity concept in the laboratory. Some of these will involve students in team research, group analysis of processes, group construction, or group investigation.

Team Research

Using groups of students doing team research has a number of advantages for learning. As a teacher, it requires you to set the stage for such an activity. An introduction and discussion about the research project should establish the objectives and motivate students to work together cooperatively. The research topic should be defined and the responsibilities for each member of the team established. Working with the team, you should serve as a resource person in assisting the plan of attack, the procedures to be used, and methods of writing a research report. Quite often a team research seminar provides an opportunity for students to evaluate their study and bring together concepts they have learned to provide a total understanding.

Topics for team research should be carefully considered as to appropriateness to your course content and the ability of the team to handle the problem. Typical examples of team research topics would include:

1. To study the effects of weathering on a variety of metal finishes.
2. To study the practicality of using solar energy for cooking purposes.
3. To investigate prefabrication techniques in modern home construction.
4. To evaluate the effects of moisture on different types of plywood.
5. To study the use of plastics in the manufacture of automobiles.

Other laboratory activities involving group learning may be an analysis or technical study that requires the use of instruments and study guides or charts to accomplish the desired learning. Most common of activities of this nature are keys or charts for material identification and structure, design procedures, and diagnosis techniques. These can often be studied through reading or lectures; however, the actual "hands on" experience of groups using an automotive engine analyzer or following a check list in locating a short in an electronic circuit provides greater retention and understanding.

The possibilities of student questioning, discussions with each other as the activity progresses, and involvement in evaluation of each others' work reinforces the learning potential. With this type of group activity, the teacher should take the lead in getting things started, explaining techniques, answering questions, and then slowly turn the complete activity over to the group. He then assumes the role of observing the group to make sure they are using the equipment correctly, posing questions that the group may overlook, and encouraging individuals to rely on each other as learning proceeds.

An outline to assist the group in their planning the activity may make use of any or all of the following:

1. Purpose.
2. Materials used or tested.
3. Instruments, tools, equipment, testing machines.
4. Procedure.
5. Observations.
6. Conclusions.
7. Sources of error.
8. References consulted.

RESEARCH REPORT
TEAM MEMBERS

Mass Production

Another group activity that may bring about the desired behavior changes for your technical course is that of a mass production unit, or a line production experience. This type of activity is desirable in meeting a number of course objectives. In some instances it gives the student an opportunity to make an interpretation of industry. In others, it provides a means for the production of many needed products, while the learning emphasis may be on specific skill development in making jigs and fixtures or machine tool setup. Teachers sometimes avoid using a line production activity because they believe it is too difficult to organize or that the outcomes may not be satisfactory. Organization is the key to this type of laboratory activity. There are many ways to approach the problem. Perhaps the best way to illustrate an organizational pattern is to outline a typical line production experience where the teacher's role is primarily that of a resource person, educational, and technical consultant. Of prime importance is the class time spent with the teacher in preparation of the group, with explanations and discussions, so they are ready to begin.

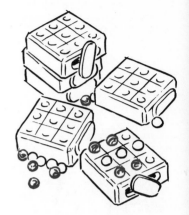

The following suggestions for a typical line production organizational pattern may be adapted to a variety of products:

1. Use your class discussions and student participation to form an "executive board."
2. With the executive board taking the leadership, evaluation of the type of product desired should be discussed. It may be necessary for you to guide and assist on the final decision as to the scope and limitations of the product.

3. After the product has been identified, present explanations which will lead the group to the formation of a company organization. Your facilities, the product, and course content will help determine the type of organization to be used.

4. Typical organization from the executive board may include the following departments: Cost Analysis - Industrial Relations - Safety Department - Production Control - Engineering - Tool and Die Department - Manufacturing - Maintenance - Sales.

5. Responsibilities growing out of these departments may include designing, jigs and fixtures, models, tooling, operations, machines, and clean up.

6. Each department should present a report to the executive board as their phase of planning progresses.

7. When all planning is completed, a "flow chart" for production should be prepared.

8. After production is under way, progress reports from each department should be made to the executive board for evaluation. Necessary changes growing out of these reports can then be made in production.

9. When production has been completed, each department should make a summary report. This would include such items as sales statistics, time and motion study, job evaluation of students, and suggested methods of training workers for industry.

10. A group seminar should be held to evaluate the total learning situation as related to industrial production.

When a unit on mass production appears to bring about a better understanding of your course content and the established objectives, make use of this valuable learning technique. Variations in the suggested mass production outline may be made to provide exciting learning experiences in many technical courses, from aesthetics to synthetics.

Much attention has been given to the planning and development of manipulative learning activities since it appears this has been one of the real needs of the industrial education teacher. You should be well aware that the activities outlined have covered a broad spectrum of student and teacher involvement. Review once again the concept that the more varied you make your laboratory activities, the more opportunity you are giving the student for optimum learning.

These activities have been directly involved with content derived from your course outline. There are certainly other activities that should be considered that you may develop to supplement the content. Add these to your list, and then in Chapter 6 you will have an opportunity to bring them all together in lesson planning and presentations.

Activities That Supplement The Content

The teacher is sometimes so preoccupied with the details of technical course material that he does not give enough attention to activities that are supplementary to the content. You should also concern yourself with activities that are related to human growth and development within the perspective of your course. Such activities should be related to course content and aid the student in better understanding his role and abilities in the learning process. These are usually "built in" to planned activities and are not separated segments. A review of the laboratory activities already discussed will show how many of these have placed the student in a position to do his own planning, research, problem solving or analysis. In many cases, these planned student approaches to laboratory activities are just as important as the technical content within.

Learning How to Learn

Learning to earn has a certain "jingle" that motivates and excites the learner. Learning how to learn sounds boring and dull. Yet, the latter is just as valuable or even more so, than the first. It denotes an ongoing process of mental development that a student acquires and carries with him through life. In some instances it is not a well developed or consistent mental ability. In others, it is, and it provides the learner with the mental tools for handling and solving problems with ease.

You have undoubtedly observed this quality in many people. They do not have to be pushed and prodded to tackle an assignment. They do not keep asking the teacher questions, nor do they keep looking around to see how others have approached the problem. Invariably they go right ahead pursuing the assigned task and come up with a satisfactory solution. Certainly this whole concept deals with the psychology of learning and mental development which is not the central theme of this discussion. However, you should be concerned that your students are developing good mental habits of learning. In other words, activities that cause the student only to memorize, copy, or imitate are not enough. They should be planned in such a manner that they require certain mental approaches to solving problems. Certainly you are not going to say to the class "today we are going to learn how to learn." On the other hand, you may well be doing just that without saying it. For as you introduce an activity that requires, perhaps, reflective thinking to provide a satisfactory solution, you are helping students develop good mental habits of thinking and learning.

Mental skills acquired in conjunction with the study of the subject matter do not cease at the conclusion of the course. They become part of the students mental approach to learning, which should be reinforced in every course. Then, outside of formal courses, he may be well equipped to solve technical problems with a higher probability of success. Illustrated activities which cause the student to define a problem, state tentative solutions, collect data, do research, propose a solution, carry out the activity, and evaluate, are vital links between content and learning how to learn.

Discovering New Interests and Insights

In any technical course you have the opportunity to assist the student in discovering new interests and insights in specific areas or broad fields of industry. Probably no other curriculum area is in as good a position to provide this potential as industrial education. It is overwhelming to think of the possible interests to which you have the opportunity to expose your students. Depending upon the technical area in which you may be teaching, you should provide avenues for the student to become involved with and interested in many aspects of industry, science, and technology. Exposure is probably the significant factor in discovering new interests on the part of the student. How can anyone get enthused about something he knows little or nothing about?

Most students know very little about the many facets of the wide world of industry. Each laboratory activity, unit of study, field trip, or presentation should be organized in such a manner that it will allow the possibility of further investigation. It is in the informal atmosphere of the industrial education laboratory that you have the opportunity to feel the response of individual students to new activities, and hence, provide meaningful resources which will further whet their newly discovered interests. Encouragement of individual students to further pursue a new interest may well have an important inpact on their lives.

Newly discovered interests may be of a hobby, recreational, scientific, occupational, or professional nature. When an inquiring interest is observed, you should make every effort to help further the student's investigation or study. One of the most important aspects of discovering new interests is the possibility that it may lead the student to an awareness of occupational or career opportunities. As an industrial education teacher, you have the responsibility of assisting the student when he shows an interest in some occupational field.

Occupational Opportunities and Requirements

Obtaining and holding a job, whether it is as a skilled worker, a technician, a scientist, or as a professional worker in industry, is a complicated task if one takes an intelligent approach to the situation. Too often students leave school to enter the world of work with little or no experience or knowledge of the occupational situation, or how to make a serious attempt toward employment. Although most programs in industrial education do not have specific courses in occupational opportunities or vocational guidance, each teacher within the program should be in a position to assist students in preparing himself for employment. Class discussions of job opportunities, educational preparation, writing letters of application, performance requirements, unions, professional associations, and job advancement are a few of the types of topics of value in the technical course you may be teaching.

For example, if you were teaching a course in electricity-electronics, there would be a direct connection between most of the activities in which your students were engaged and similar situations going on in industry. You could use many illustrations to inform the student that the tasks he is performing, or equipment he is testing, are being done, perhaps at an advanced level, in similar industrial situations. You could point out the necessary background required for specific jobs, such as a television repairman, electrical engineer, computer designer, and others. It would certainly be interesting to know how many people are engaged directly or indirectly in occupations in the electronics industry who were motivated or inspired by such courses in their educational program. It is your responsibility to provide an introduction to occupations for your students. The use of instructional activities that provide for a better understanding of employment opportunities and occupational requirements often serves this purpose. This type of individual study activity usually provides more insight for student understanding of industrial occupations than other methods. He is active and does his own investigation and evaluation. An example of some of the topics you may use in developing an occupational study activity sheet are as follows:

1. Develop a title for your particular occupational study outline.

2. Provide sources of occupational information for the student to investigate. These may include occupational handbooks, dictionaries of occupations, guides and handbooks for workers, reference books, and local and state employment services.

3. Develop an outline for the student to follow in selecting an occupation or occupations to study. This outline should include statements and questions to aid the student in his search for information and understanding.

4. Topics for study of the student selected occupation may include:

Future Prospects - Is employment expected to increase or decrease?

Nature of Work - Define the duties of the worker.

Qualifications - Upper and lower age limits, height, weight and other physical requirements.

Legal Requirements - Is a certificate or license required?

Preparation - Education requirements, length of time, cost, on the job training, employer's standards.

Entrance - How does one get his first job? By application, examination, employment agencies, unions?

Advancement - What requirements are necessary for advancement, what proportion of workers advance?

Earnings - Comparison of earnings upon entrance with other jobs. Upper limits for earnings.

Advantages and Disadvantages - What workers like or dislike most about their job. Is the employment steady? Frequent overtime or night work?

Related Occupations - Other occupations to which this job might lead. Skills that are transferable.

Review Questions - Chapter 3

1. How has the psychology of teaching and learning changed over the years on the part of the teacher and student?
2. What learning concepts does a line production experience provide for the student?
3. What two major factors should be emphasized in planning an activity for "operating instructions" for a specific piece of equipment?
4. Explain why a broad spectrum of laboratory learning activities, from teacher designed to student planned, provides a greater potential for learning than one specific type.
5. What is the difference between learning activities that are referred to as logical or psychological?
6. What are the major features of a well planned learning activity?
7. Why are activities that supplement the content of a course so important to the learning situation?
8. What is meant by learning how to learn and why should laboratory activities emphasize this principle?
9. How can the teacher implement activities that will help the student gain new interests and insights concerning the technical area being studied?
10. Describe the fundamental principles of an activity that is designed to create a problem solving situation.
11. What topics are of major importance in reference to job opportunities and requirements?
12. How should a teacher go about selecting laboratory activities that will best meet the needs of his students?
13. Explain how passive learning differs from activity learning.
14. How does sense perception relate to learning?
15. What are some of the principles of learning that can be obtained from a problem activity?
16. Why is it a good idea to have students use a prepared planning activity sheet when he is going to do his own planning?
17. What are some of the major values of group learning activities for the individual student?

Suggested Activities

1. For your particular technical area of interest, develop an activity sheet for the correct operation of a new piece of equipment.
2. Make a brief topical outline of a line production activity that would be suitable for your technical area.
3. Using reference materials and library resources, write a paper tracing the changes that have taken place in assigned laboratory activities over the past forty years. Locate and explain the meaning of information sheets, assignment sheets, job assignment sheets, operation sheets, and information assignment sheets.

4. Design a testing activity to be performed in the laboratory in your technical field. Use an industrial standard as the basis for setting up and evaluating the testing procedure.

5. Make a chart that indicates the general requirements for a skilled, technical, or professional occupation in your technical field. Include aspects for preparation, successful performance, and in-service education for the occupation selected.

6. Design a product, specifications, procedure, and evaluation for a laboratory activity to bring about the desired behavior changes desired for a particular objective of your technical course.

Chapter 4
THE INDUSTRIAL EDUCATION LABORATORY

The type of laboratory best suited for courses in industrial education has long been a topic of discussion among teachers. Well it should be, for there is no one general purpose laboratory that meets the needs of the great variety of technical courses offered in our schools. However, there are some common considerations that apply to the planning and layout of all technical laboratories. Let's look at this list:

1. The industrial education laboratory should be contemporary in nature.
2. The laboratory should be developed to meet the specific needs of your course or curriculum planning.
3. It should reflect the spirit of industry for whatever technical course is planned.
4. It should be designed for the greatest ease of teaching and learning.
5. The laboratory should contain as much up-to-date equipment as may be purchased within the school budget.

A more detailed look at these common features will give you a better overall view of modern laboratory planning. Certainly a newly constructed laboratory makes it easier to plan on a contemporary basis. Even so, an older laboratory may be brought up to date by redesigning and rearrangement. The use of the term contemporary refers to the overall layout of the laboratory, such as lighting, work and planning areas, location of equipment, storage, and colors, to mention a few. The well planned technical laboratory will be contemporary in nature because it will represent the futuristic philosophy of the particular area of industrial education for which it is to be used. Plan and keep your laboratory as contemporary as possible.

Your course plan should be the determining factor in laboratory planning. Since your course plan presents the material you are planning to teach, and your laboratory activities have been developed, the nature of your laboratory should be such that these activities may be feasibly carried out. Think of trying to teach a graphic arts course in a metals laboratory. An extreme example, but you get the idea, the laboratory is not suitable for the planned content. Realistically, the laboratory should be set up in such a way that all planned activities may be performed by the student with little inconvenience and reorganization during study.

Any technical laboratory should reflect the concepts and spirit of an industrial setting. Reference is again made to the relationship of industrial education to industry. If your course content reflects modern industrial concepts, so should your laboratory. Visits to industries which represent your technical area of teaching will often give you good ideas for laboratory arrangement. Such questions as: What did their production area look like? How was their research area arranged? Where did they do testing and analysis? How was equipment arranged? will aid in your planning.

Along with the desired content and activities to be carried out in your laboratory, is the need for the laboratory to be arranged for ease of teaching and learning. Plans should be made for adequate space for group activities, multimedia presentations, planning center, reference materials, flow of students through activity areas, independent study and research. It is not easy to plan a laboratory for optimum learning. In most cases you must teach in your laboratory for a considerable time to spot the points of difficulty. As difficulties arise, adjustments can be made. The continuous process of laboratory arrangement allows for the best in teaching and learning.

Another common feature of all technical laboratories is that they should contain the latest equipment and supplies available. This sounds good and would work well if you had all the money necessary to secure such equipment. The point is, keep your laboratory up to date within your budget, and take advantage of any donations you can secure from industry. Old, worn-out equipment not only deprives the student of gaining modern industrial principles but also gives an obsolete, uninviting appearance to your laboratory. Painting, reconditioning, trade-ins, new purchases, and donations can go a long way in updating your laboratory. A 1918 model giant belt-drive drill press may look impressive in a museum, but not in your modern laboratory.

Your Laboratory - Your Responsibility

The responsibility for maintaining and continually upgrading the industrial education laboratory is usually placed in the hands of the individual teacher. This responsibility should be accepted with pride. The laboratory is your teaching center, and your enthusiasm should build up as you devise methods and equipment for making it a better place in which to teach and to learn.

Space limitations prevent us from providing here, a detailed discussion of individual laboratories for the many technical courses, and for the levels at which industrial education should be approached. Numerous references are available for laboratory planning for specific technical areas. Review these resources for specific laboratory development for the technical area of your interest. In this Chapter we will discuss your responsibilities as a teacher in the laboratory in general terms.

Tools, Machines, and Equipment

A problem that will always be with you in technical teaching is that of ordering, maintaining, and care of tools and equipment in your laboratory. Equipment in any laboratory should always be in good condition and ready to be safely used. Many teachers become bogged down with the burden of maintenance and repair until their teaching program suffers because of little time left for planning and organization of teaching materials. You must budget your time in regard to equipment maintenance. The industrial education teacher is employed first as a teacher, and care of equipment is just one part of that job.

A desirable solution to this problem is the employment of a technician for the total school program who can be called upon for repair of equipment. Some schools make provision for employing such a person in their budget.

Another solution is to incorporate some maintenance and repair of equipment in your course planning. You should be careful in this respect to be sure it becomes a meaningful learning experience for the student and falls within the objectives of your course. An example might be that a grinder has a worn-out wheel. A demonstration of replacing the wheel and truing the grinding surface may be a vital learning experience. It may also save the teacher the time of coming in at night do the same job. A careful evaluation should be made for any such situation in line with your course content and objectives.

Sometimes student involvement in equipment repair is appropriate, if it is within the scope of the course. Repair of electronic equipment in an electricity course may well fulfill the goals of appropriate learning activities in that area. In the same course, having the students paint storage cabinets would be of little interest.

Another point is to minimize repairs by placing heavy emphasis on the correct use of equipment during your laboratory teaching. Equipment and tools are often damaged by incorrect student use. Many hours have been required for repair and maintenance when students have tried to cut hardened tool steel on the band saw, surfaced boards containing nails, cut heavy bar stock on the squaring shear, connected up an ammeter incorrectly, or let the ink dry when doing a silk screening. Problems such as these can be avoided if the proper instruction, attitude, and observation are employed.

Purchasing Equipment

Once again your course plan and desired laboratory activities should be used as a guide for the purchase of equipment. It may be tempting to some teachers to use a selected list of equipment from a catalog, textbook, or another teacher. Certainly it is quicker than preparing your own specifications, but it may not include items needed for your specific laboratory and teaching plans.

When you are faced with the problem of preparing a list of equipment desired for your technical laboratory, it is advisable to determine exactly what you need and how many of each item. This may be determined by the activities you have planned and how many students may need to use certain equipment at the same time. You must also consider your available space and the number of students expected in each class.

With these things in mind, it is a good idea to place a priority on items on your list consistent with an estimate of your budget. The priority should be determined by your immediate needs for those planned activities which will help your program most. Purchasing equipment is an ongoing process. It is doubtful if any technical teacher ever had what he considered a completely equipped laboratory, for two reasons: First, budgets for most programs are usually not sufficient to allow purchasing every item desired. Second, the process of planning new activities, and new equipment becoming available on the market, makes it necessary to keep your list active. Before sending in orders, it is a good idea to check out some pieces of equipment to be sure you will be satisfied with their quality and performance. One method is to have the distributor or sales representative demonstrate the equipment. Another is to discuss various equipment with teachers who have those items and can give you viewpoints from experience.

Some suggestions for purchasing equipment:

1. Take time to go through catalogs and manufacturer's specifications to be sure each piece of equipment will perform as expected and is up to date.

2. Investigate all standard makes and select equipment that will meet your specific needs. Your laboratory should not be an advertising display for any particular manufacturer.

3. Be sure you communicate clearly with the supplier by making your specifications accurate.

4. Use a standard format in preparing your purchase order or requisition. This should include the manufacturer's name, item number, quantity desired, model number, name or description of the equipment and unusual specifications.

5. Keep an accurate inventory of all laboratory equipment.

During courses you are or have taken, make an effort to get acquainted with all aspects of the equipment in the laboratory. There is just as much to know about most machines and equipment as there is to using them.

Your Teaching and Planning Area

Your teaching and planning areas are an integral part of the industrial education laboratory. Most educators no longer think of the laboratory as a place where students just go to work. The contemporary view is that the technical laboratory is a learning center and the days have passed when classroom study and work were separated. Today the technical requirements of industry dictate the need for laboratories in the schools to integrate areas for experimentation, testing, planning, multimedia learning, construction, production machine operation, research, and communications. Although the many technical areas of study will vary considerably, from such extremes as drawing to foundry work, each should take advantage of the latest planning for teaching and learning activity areas. In other words, the whole concept of learning should override traditional placement of equipment and desks. If a table for microscopic study of the fracture of metal samples provides the best setting for learning when located next to the pouring area, there it should be.

The teaching and planning area should include materials and equipment necessary to provide an optimum learning atmosphere for you and your students. Much of this is discussed in detail under the sections on reference materials and instructional media. For here it should suffice to suggest the concept that all desired teaching and learning situations be considered in designing a learning center. Perhaps an example will help to clarify the point.

Place yourself in the position of a graphic arts teacher and consider the layout of a laboratory to meet the needs of both the contemporary technology and psychology of learning. You would undoubtedly need facilities for students to do planning and layout work, group discussions and presentations, testing and experimentation with colors, papers, inks, and a multitude of other materials, operation of all types of graphic reproduction equipment, photographic and darkroom processes, and many more. The whole laboratory becomes a learning center with special facilities for student planning and design.

Any industrial area, whether in technical, vocational, or industrial arts education, should provide facilities for student planning, study, research, and design activities to some degree. The objectives of the individual programs will give direction to how much emphasis should be placed on planning and student study areas.

Supplies - Ordering and Storage

Well thought out and planned procedures for ordering and storage minimize the time spent on this part of your laboratory responsibility. Most schools have an established system for ordering supplies on standard requisition forms. If your school does not, you should develop one of your own. In any case, you should make duplicate copies of all supply requisitions for your own file. This duplicate form makes it possible for you to check off supplies received, those back ordered, and those not available. It also gives you a permanent record of supplies purchased, amounts and costs. Reference is often made to earlier orders in planning for future courses.

An ample stock of supplies should be maintained for all planned and predicted activities. When ordering, you should specify on your requisition form exactly what you desire. According to suppliers, it is amazing how many problems are encountered because teachers do not specify the exact size, color, quantity and so forth.

Check incoming shipments carefully and report any errors or shortages.

Any well-organized laboratory will provide for a systematic storage system. It is important that you provide a systematic arrangement that will meet your particular needs. Use reference materials and magazine articles for ideas and suggestions for your particular area, then plan storage space, cabinets, and containers for all supplies. A system of supply storage may well work nicely for you and be unsatisfactory for another teacher. Here are a few general suggestions which may be helpful in your planning:

1. Check industrial catalogs for suitable storage racks and cabinets. Available equipment may suit your needs.

2. Visit and talk with other teachers in your technical area. Some may have solved difficult storage problems and can give you some good ideas.

3. Provide plenty of storage space since new materials and supplies may require additional room.

4. Make your supplies accessible. In some cases supply cabinets may be open for student use; in others you may keep certain supplies in locked cabinets for your own personal distribution.

5. Storage of some items may require specific conditions to maintain shelf-life or quality. Photographic or plastics chemicals may need refrigeration, certain materials must be kept out of daylight, etc. Check the requirements on such supplies and provide the proper conditions.

6. Identify supplies by name or number in storage cabinets or containers. Any that are toxic or dangerous, should be clearly marked and kept in locked cabinets.

7. Storage rooms separate from your laboratory are often advantageous so you can control the use of supplies and easily check your inventory.

8. Consider the location for storage of large or heavy materials which may have to be unloaded from a truck. If possible, these should be located close to the unloading area.

The better your storage arrangements for materials and supplies, the easier your teaching situation. Do not be a material chaser. Have supplies at your fingertips.

Your Display Can Be Vital

Few people will know what is going on in your laboratory unless you show them. Some may have preconceived notions which do not properly reflect your program. Give students, teachers, and the public an opportunity to see and understand the learning activities in which your students are engaged.

Displays may take place in a number of forms. You should have a display area within the laboratory. This provides an opportunity for visitors and students to see the processes or end results of the many learning activities taking place in your courses. It is often difficult to describe the purposes of your technical laboratory to a visitor who views only a mass of scientific instruments, technical machinery, or production equipment. A display of the processes, design activities, research reports, products, and industrial pictures may well put the point across quickly, in an interesting and informative manner.

A visitor in a graphics laboratory who views an offset printing unit or an automatic silk screen printing machine may be interested in the basic processes of printing. A display of the processes involved and the resulting products will give him a good insight into the content of your program. Such displays may also provide the visitor with topics of discussion he will convey to others in the community. This is vital to an understanding of your program.

You should also provide displays for hallway or lobby display cases in your school. These displays may depict various aspects of your program or related material secured from industrial organizations. Many industries are willing to provide products, pictures, sequence processes, and even some equipment on a loan basis for display purposes.

Another avenue for communication of your technical content is the use of bulletin boards. Attractive bulletin boards illustrating student activities or industrial practices brighten up your laboratory and convey appropriate messages. Students may work with you or do their own planning and designing of displays. In many cases worthwhile learning activities are developed, especially if the display deals with content in which they are engaged.

Displays will reach only a limited number of people. Additional publicity may be obtained by photographing displays, bulletin boards, and other activities and having the photos printed in school and local newspapers along with descriptive articles. You may also develop articles with pictures describing new and valuable activities and submit them to professional magazines for publication.

Your Laboratory and Content Work Together

Your laboratory should be far more than a workshop with an attached classroom, it should be a fantastic learning center. This is the reason your content and laboratory should work together to promote the ultimate in student learning. As has already been indicated but cannot be overemphasized, your course content outline provides the map and specifications for equipping your laboratory. Not only with tools and machines and all other equipment including their location, but also with every conceivable teaching and learning media. When a contemporary approach to laboratory planning is used it will necessitate that your course outline and planned activities dictate the arrangement of the complete learning center.

Planning Your Laboratory for Learning

The development of an industrial education laboratory for any specific technical area is an ongoing process. It is much like planning the arrangement of furniture and appliances in a new home for the most comfortable and efficient form of living. People say a home must be lived in before they can decide exactly how things should be arranged for convenience and comfort. Your laboratory has to be "learned-in" before you can make decisions as to the arrangement of equipment, student planning areas, reference materials, audiovisual materials, and storage of supplies.

After an initial, thought-out arrangement of your laboratory, probably the best way for you to evaluate its effectiveness for learning is by careful observation of the learning process during courses being taught. You may find congested areas where students are too crowded to perform desired activities. Another common observation is that students will be waiting to make use of necessary equipment, either because not enough equipment is provided or your activities have not been planned to make necessary facilities available at the right times. Planning your laboratory for learning requires that you constantly evaluate the learning process in regard to facilities and make adjustments and rearrangements as necessary. Instructional media is included in this concept of laboratory learning. A detailed discussion of teaching media is presented in Chapter 10.

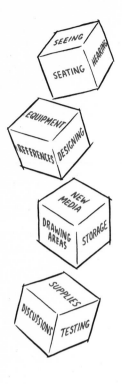

Some suggestions for laboratory planning and arrangement to provide for effective learning are:

1. Make certain that facilities for demonstrations are readily available and all students are able to see and hear.

2. Provide adequate seating and planning space for ease of study, designing, or drawing.

3. Make reference materials readily available and convenient for student use.

4. Plan equipment arrangement so each item is easily accessible for discussions, demonstrations, and individual use.

5. Make your laboratory look as if it actually represents the content of the courses you are teaching.

6. Plan possible seating arrangements for group discussions, conferences, and seminars.

7. Provide adequate and convenient storage for student supplies and products.

8. Plan instructional media equipment for maximum use and effectiveness. Correct location of projectors, screens, learning carrels, tape recorders, closed circuit television, and other equipment deserves close attention.

Reference and Resource Library

One of the most beneficial aspects of any technical laboratory is the availability of reference and resource materials for student use. It is difficult to comprehend that any contemporary technical course could be effectively taught by using no other resource material than a textbook, regardless of the quality of the text.

A set of bookshelves or magazine racks should be suitably located for student use during laboratory instruction, free periods, and after school hours. These materials will be needed often to supplement course content, provide for research activities, indicate suppliers so students may write for specifications and information, provide for advanced investigation, and instructions for equipment operation and testing. Materials you should consider for references include technical books, manufacturers publications, technical association magazines, industrial literature, professional magazines, career publications, and teacher prepared references. Some feel that these materials should all be kept in the school library, however, many teachers find that students do not use a school library efficiently. Frequently information such as material specifications, is needed immediately - - not when the student has time to go to the library. Just as you reach for a reference magazine on your desk to plan an activity, the student should be in a position to do the same in your laboratory.

In order to keep reference materials up to date and available to all students, a sign over your resource center indicating, "FOR LABORATORY USE ONLY," would be quite in order.

Does Your Laboratory Reflect Industry?

Visit a modern local industry and get the feel of the environment; the offices of management, the design and engineering departments, research and development, material and product testing areas, reference libraries, manufacturing, production, packaging, sales department. Then go back to your industrial education laboratory and begin to make comparisons. You are not developing a manufacturing or service company, but does your laboratory reflect industry? After all, industry is the total framework from which your subject matter is obtained, and it is where many of your students are headed. You have obtained your course content from the world of industry, now place that content and your students in an industrial setting. LET YOUR LABORATORY REFLECT INDUSTRY.

The concept of an industrial education laboratory that resembles the industrial setting provides the student with a realistic transition from learning to earning. A field trip to an industrial concern will often surprise students, not so much in what is being manufactured or processed, but that it does not resemble the same content they have been studying in their own laboratory. Certainly you cannot have all of the same equipment used in industry. Most of it is much too expensive, too large, and on a production scale far beyond your needs. However, similar equipment on a smaller scale and a variety of instructional media will introduce your students to the same learning concepts. So plan your laboratory along the same principles that industrial planners have used.

Take Advantage of Color

For many years experts in the field of color analysis have stated that certain colors for walls, machinery, equipment, and other facilities place less eye strain on workers and provide for more efficiency and safety. Your student is in a similar position, and you should take advantage of painting and decorating techniques that will produce a laboratory with minimum glare, a soft texture of surroundings, and pleasant working conditions. Check with the manufacturers of paints who specialize in color decorating for industry. These companies will supply you with instructions in the use of color dynamics which have been successfully used in industry.

Complete industrial color systems not only provide schemes for better worker efficiency and better eyesight but also for safety purposes. Standard color codes are available for marking pipes containing gases and liquids, danger areas on and around machinery, stairways, heating devices, and electrical panels. Take advantage of available color systems. If an overhanging beam upon which students may bump their heads says "color me yellow," try it; it may save more than bruises.

Take Advantage of Lighting

Natural light or sunlight is desirable for most laboratory conditions. However, in many instances not enough natural light is available. You should consider each of the working and study areas in your laboratory for necessary light conditions. Most schools provide adequate interior lighting for general purposes. It is the specific areas of activity that you should provide lighting that will produce a minimum of eye strain. Detailed work such as drawing and material analysis require a higher degree of lighting than do most other areas. Shadows should be avoided around equipment where students are doing precision work. Lighting conditions should also be analyzed when using many of the audiovisual materials. When using overhead transparencies, for example, enough light should be available for students to take notes and see the screen without eye strain. An average of 60 to 100 foot candles of light for general laboratory purposes should serve as a satisfactory guide. Specific conditions will require your own good judgment along with suggestions from industrial personnel who have solved similar problems.

Provide Adequate Storage

In many instances adequate space is not provided in the laboratory for the storage of materials or products in various stages of construction or testing. Whatever the laboratory activity may be, a line production unit, a machine product, or materials being used in an ongoing experiment, you should have space available where these materials may be stored until the next class meeting. Specified shelving or a storage room will usually meet these needs. Such material and equipment should always be put away when the class session ends and not be allowed to lie in the laboratory. Exceptions, of course, would be certain equipment setups which cannot be moved until the process is completed. Student lockers will accommodate a variety of small items.

Review Questions - Chapter 4

1. What are the common features that apply to the planning and layout of most all contemporary laboratories? Can you list any others?
2. What is the major determining factor in laboratory planning for any technical area?
3. How can the many problems of equipment maintenance and repair be minimized on the part of the teacher?
4. Where can you locate references for laboratory planning of a contemporary nature for your technical area?
5. What factors should you consider when making up a list for ordering equipment?
6. Why is it a good idea to keep a file of your own on supplies ordered?
7. In what ways can your displays play an important part in your technical teaching program?
8. Explain how your laboratory and your course content should compliment each other?
9. What are some advantages of having a reference and resource library in your laboratory? What types of materials should you make available to students in such a library?
10. In what ways does color influence your laboratory as a teaching-learning center?
11. Why is it so important to relate technical laboratory planning to concepts of industrial working conditions?
12. How does proper lighting pertain to good learning activity centers in the laboratory?
13. What are the advantages of having proper storage areas for supplies and equipment that students are using during learning activities?

Suggested Activities

1. Make a visit to an industrial concern dealing with your particular technical field and discuss the problems of tool storage, machine maintenance, equipment arrangement, working conditions, and material storage. Draw up a list of factors that would be helpful in planning your laboratory, from the information you have received.

2. Write to a number of paint companies that specialize in industrial color dynamics and ask for literature on solving laboratory color problems. From the literature you receive, devise a color scheme for your laboratory planning that takes advantage of the research that has been done on the effect of color and working conditions.

3. For your technical area, make a floor plan drawing for a laboratory using the principles of contemporary laboratory planning. Consult with other industrial education teachers, industrial personnel, and modern reference materials before beginning your plan.

4. For your technical area; make a listing of any special lighting conditions which would be necessary in your laboratory for ideal conditions. Consider such factors for lighting as needed for reading dials on test equipment, microscopic viewing, equipment operation, planning centers, and storage areas.

5. Make a list of the latest in laboratory equipment that would be necessary to teach the concepts drawn from your course outline and planned learning activities. Check as many sources as possible to be sure the equipment is dependable, practical, modern, and designed for maximum student learning and use.

6. Begin a collection of supply catalogs dealing with your technical specialization. Make a survey of industrial suppliers as well as those who specialize in supplies for industrial education. Talk with other teachers and people from industry who may give you tips on supply sources. Many industries will donate or make available at reasonable cost, equipment and supplies for educational purposes. Include such possible sources in your collection.

Chapter 5
CONCEPTS OF SAFETY
AND FIRST AID

In the school shop, you cannot overemphasize the importance of safety rules and practices. If you are demonstrating or instructing the use of equipment to your students, show them the reasons for correct safety practices. Give them an opportunity to completely understand how accidents may happen if equipment is used incorrectly.

A SAFETY ATTITUDE MUST BE DEVELOPED IN STUDENTS TO FOCUS THEIR ATTENTION ON THE PREVENTION OF POSSIBLE ACCIDENTS. Students should develop concepts of safe work habits and an awareness of the dangers that exist when handling equipment, tools, and machinery. You should emphasize both the physical and human elements that cause accidents. The physical elements can be stressed by frequently calling the student's attention to the dangers associated with all the equipment and processes done in your laboratory. The human elements which cause accidents are more difficult to handle. These include such characteristics as laziness, ignorance and carelessness.

There are many common sense precautions which you, as a technical teacher, can practice and insist that your students observe. It is your responsibility to assist the student in developing safe work habits and a proper respect for the hazards of working with technical equipment.

Safety is not a matter of a unit to be taught during each course. SAFETY SHOULD BE AN INTEGRAL PART OF YOUR CONTENT FOR EVERY COURSE. Not just rules and regulations, although these are important, but the development of a positive mental attitude toward safety on the part of the student. The development of such attitudes hinges much on your attitude toward safe practices and your ability to plan activities to bring them about.

To begin planning for safety in any school laboratory, there are a number of concepts to be considered. They may be classified under these headings:

Preparing Your Laboratory

One of the first concepts toward safety is the laboratory itself. You should constantly inspect your laboratory for possible hazards which may cause injuries that may be the fault of poor equipment arrangement or unsatisfactory preventive measures.

Space is not available to review all possible preventive measures. However, a few examples of common potential dangers that often exist in laboratories should serve as a guide when you are planning to make your facility the safest possible for student activities:

1. Floors should be made as nonskid as possible by using some of the newer waxes or finishes.
2. Any objects that extend into passageways where students may bump themselves or trip should be corrected.
3. Materials or units that are hot, electrically, or mechanically dangerous, toxic, or dangerous to the eyes should be so marked.
4. The necessity for using ladders or chairs to reach materials and supplies should be eliminated.
5. Slippery stairs in the laboratory should be covered with slip-proof treads.
6. Areas where chemicals or smoke are emitted should have exhaust systems.
7. Equipment should be arranged so students have room to work without bumping other equipment.
8. Eliminate all trash, unused scraps, equipment, and containers that may be a potential for accidents.

It has been estimated that up to 15 percent of all laboratory accidents would have been eliminated if the facilities were properly prepared before student use. Laboratory preparation is more than an inspection at the beginning of your courses. It must be an ongoing process throughout all of your teaching activities. This includes your responsibility for student cleanup and inspection of the work areas after each session. Such things as leaving the laboratory with the gas from a bunsen burner or oxyacetylene torch left on, a bottle of acid not capped, a heating oven not turned off, or hot electrical wires exposed, invites possible injury to those who may enter the laboratory at a later time.

Physical Elements of Safety

Safety in technical work areas in industrial education depends heavily on instruction in the physical use of materials, dangerous solutions or chemicals, heating devices, electrical devices, mechanical equipment, and testing equipment. These materials and equipment can be used safely if proper instruction is presented in a positive manner.

One of the best concepts of safety lies in the positive attitude of "DON'T TRY IT IF YOU'RE NOT SURE IT IS SAFE." Try teaching this attitude to your students. Students often feel they can "get away" with doing things in an unsafe manner, and sometimes they do. However, serious accidents may result when they do not. Another good concept to instill in your students minds is, "IF THERE APPEARS TO BE ANY CHANCE I MAY GET INJURED BY DOING SOMETHING A CERTAIN WAY, DON'T DO IT." Find another way that may involve less risk.

Physical elements of safe work practices involve the use of good judgement, protective equipment, and a knowledge of how the equipment operates. Let's consider a few examples of the physical elements that relate to accident prevention. These examples, representing many types of technical shops and laboratories, are used to illustrate this concept of safety. For a specific industrial education laboratory specifications from resource material and manufacturers literature will be of particular value to you. Some equipment is inherently dangerous. By calling attention to these dangers you can provide the student with definite concepts of safety due to physical elements:

1. Eye protection - Safety glasses and other devices should be used at all times where the possibility of eye injury exists. Most injuries to eyes come from flying objects, intense light, splashed liquids, or high heat. These include chips from cutting devices, bright light from welding operations, explosions from heated liquids or acids, sparks from an incorrectly connected electrical circuit. Protective shields and instructions are a must in all situations of this nature.

2. Bodily protection - Physical elements for protection of the body are of special importance. Hands are particulary vulnerable. You should teach the concepts of safety by the use of protective equipment and rules of operation. TRY TO INSTILL THE CONCEPT OF RESPECT FOR EQUIPMENT. A stamping press does not know where the student's hands may be when it closes, but the student must "know" where they should be to prevent injury. Guards, shields, and other protective equipment must be used on all equipment. Protective clothing should also be provided or required for special activities. Welding gloves, leather aprons for foundry work, and asbestos gloves for handling hot molds are typical examples.

Human Elements of Safety

SAFETY ATTITUDE IS A CONCEPT WHICH SHOULD BE TAUGHT THROUGHOUT ALL TECHNICAL LEARNING ACTIVITIES. Perhaps the best attitude you can possibly give the student an opportunity to learn is to "THINK SAFETY." If you watch well-trained workers over a period of time you notice their concept of safety has become a habit.

The human element in safe working is the most difficult attitude to acquire. It must be taught, studied, understood, observed, and practiced. The concept of safety is a human attitude to be learned just as any other educational or technical principle. It's your responsibility to see to it that the safety attitude is taught, understood, and practiced. Some suggested concepts to help you develop the conscious attitude of safety on the part of your students follow:

1. Emphasize the importance of "thinking through" the process in terms of safety before performing the process.
2. Instill the concept that "carelessness" invites injury while thoughtfulness minimizes accidents. Make it clear to your students that "horseplay" will not be permitted.
3. Teach students to be alert. Many accidents can be avoided.
4. Advise your students not to "take chances." Safety is not a game of chance where you bet on the odds. If the student feels he is taking a chance he should stop and search for another method.
5. Avoid having students perform an operation or test of which they are afraid. Apprehension or nervousness may cause accidents. Let the student become completely acquinted with the process by demonstrations and discussions until he is at ease with the equipment.
6. Be sure the student is dressed safely for the activities in which he engages. Long sleeves, neckties, loose clothing, or jewelry may get caught in equipment.
7. After demonstrations and discussions, supervise the operation of equipment by your students to make sure they understand and have mastered the safety concepts.

You can minimize the human element by:

1. Spelling out rules and regulations for use of equipment.
2. Using your bulletin board to emphasize the safety precautions you are teaching.
3. Always presenting the positive side of safety.
4. Insisting on neatness, know-how, and accuracy in the use of equipment.
5. Demonstrating correct procedures and pointing out potential dangers during instruction.
6. Checking student understanding and performance during manipulative operations.
7. Conducting class discussions involving safe procedures to be used by the whole class.

Equipment Safety

Reference here is made to the inquisitiveness of students to push buttons or turn switches to see how something works.

You should acquint your students with equipment in such a manner that they will not try something just to see how it works. Much of the equipment in modern industrial education laboratories can be easily damaged. Dropping the arm on an impact testing machine just for fun; pressing the "on" switch on a plastics extruder when it is cold just to see what will happen; or plugging in a machine marked, "out of order," to see if it really is, may cause serious damage to equipment and student alike. Take the time necessary to inform students not to touch or try to operate equipment until they have had the necessary instruction.

First Aid in the Laboratory

The normal responsibility of the industrial education teacher concerning first aid should come directly from the administration or department chairman. School policies have usually been established and you should become familiar with these as quickly as possible. As a teacher, you should study the basic requirements of first aid, and if at all possible, take a course in first aid to give you the proper background for assisting and acting in cases of injury. You should also have an approved first aid kit. This should be periodically checked and restocked.

Since possible injuries may range from slight cuts and bruises to serious burns or bodily damage, it is important for you to know what to do for all situations that may arise. Three courses of action which may be taken include:

1. For minor injuries, treat the student from the shop first aid kit.

2. For a more serious injury, call the school nurse and notify the principal. Use first aid measures until trained personnel arrives.

3. For complicated or unusual injuries, call the fire rescue squad or an ambulance.

It should be noted that in some states and local school systems teachers are not allowed to administer first aid. The student must be sent to the school nurse.

Remember, first aid means just what it says. It includes steps necessary to render treatment to minor injury and in case of serious injury to maintain comfort, warmth, and quite until professional help arrives. All injuries should be reported to your principal or department chairman. Accident report forms from the administration should be kept readily available and telephone numbers for the school nurse, ambulance service, fire rescue squad, police, and doctors services should be at hand for use, if needed. Be sure to keep a file on accidents and injuries that occur.

Teacher Liability

The industrial education teacher assumes a degree of liability in his facilities. You should obtain information concerning teacher liability in the state in which you are employed. Be sure to obtain also information on liability from the supervisor or administrator in your school system.

Negligence is usually the most common charge that can be placed against a teacher in case of accidents. To avoid being accused of negligence, remember the following:

1. Never leave the laboratory when class is in session and students are working.

2. Be sure proper safety instruction has been given before students are allowed to use equipment.

3. Closely supervise students whom you believe may use machines or equipment incorrectly.

4. Avoid allowing students to use certain equipment who are mentally or physically handicapped or even poorly coordinated. This is a personal judgement you must make if such students are assigned to your class.

5. Be sure that all safety devices, guards, and other items of protective equipment are in working order and are being used at all times.

In some states the school board or school district is held responsible for judgements against a teacher who has been held liable for injuries to a student. In other states the teacher is responsible and must assume costs for student injuries. Check your state laws and be sure you understand them.

It is advisable to investigate personal liability insurance. Professional associations such as the American Vocational Association and the American Industrial Arts Association have low cost personal liability insurance plans which are available to teachers.

Remember the importance of a well organized laboratory, extensive safety instruction, good supervision, and insistence on the use of safety devices.

Review Questions - Chapter 5

1. What are the major causes of accidents and injuries in the industrial education laboratory.
2. Explain how the physical elements of safety in the laboratory differ from the human elements.
3. Why are safety rules and regulations not sufficient as a method of minimizing accidents in the laboratory?
4. Explain how the physical aspects of a school lab may present the potential for injuries.
5. What attitudes toward safety should the industrial education teacher try to instill in his students?
6. To what are most eye injuries attributed in a school shop facility?
7. How do safety concepts differ from safety rules?
8. In what instances should the industrial education teacher use first aid? How far should he go in first aid?
9. Explain what is meant by teacher liability. What should the beginning teacher do about liability?
10. When would the technical teacher be considered most negligent in respect to student accidents?
11. Explain how you can incorporate safe working attitudes into laboratory activities.

Suggested Activities

1. For the specific technical area in which you are interested, make a list of laboratory conditions that would minimize accidents. Include location of equipment, safety devices for specific equipment, and other technical facilities.

2. Visit a local industry and, if possible, talk with their safety engineer or other person in charge, and have him explain their concepts of safety and how they train their employees toward these concepts. Try to determine the major causes of accidents in their plant.

3. Refer to a recent copy of FIRST AID, by the American National Red Cross, and prepare a list of suggested supplies for a standard first aid kit. Make your list appropriate for a high school general industrial education facility.

4. Ask your instructor to invite a doctor, school nurse, or a member of a fire rescue squad to your class to discuss accident prevention and first aid in high school industrial education classes.

5. Have your high school students prepare illustrated safety posters to depict the correct use of machines, tools, and equipment that they will be using. Discuss with your students where they should be located to establish a more effective safety attitude.

Chapter 6
TEACHING PLANS
AND PRESENTATIONS

Your responsibility as an industrial education teacher is to help students learn. Each is being educated for a specific purpose. It is your job to give the student every opportunity to learn particular skills or to apply the principles and concepts he has learned. One of the most important aspects of this educational process consists of providing the student with concise, explicit, and definite explanations and directions as to the best way to perform or approach learning activities associated with the course. You are not a "funnel filler" who pours information into students minds, but rather a guide and resource person for student actions. There is quite a difference. As you study the principles of teaching plans and presentations, you should keep in mind that it is intended to help your present ideas about how things are done and how to apply concepts. Ideas alone are worth very little. However, when ideas are expressed as action, problem solving, or applied as manipulative skills in performance, they immediately become meaningful to the learner. It is this concept that you should keep in mind when developing your presentations and explanations. Simply expressed, it results in:

IDEAS - - PRESENTATIONS - - UNDERSTANDINGS - - ACTIONS

IDEAS
PRESENTATION
UNDERSTANDING
ACTION

The emphasis here is given to plans and presentations. These are your responsibilities in connecting the link between ideas and the resulting understandings and actions on the part of the

learner. In making effective presentations and explanations, the teacher does not merely follow a series of prescribed steps or go through a series of prescribed activities designed to lead to a good performance on the part of the teacher. Instead, you should use the skill and education you have developed in guiding the developmental learning process that takes place in the student. The result of such a learning situation leads to accomplishing the objective of student behavior change where he can begin to do the activity or to apply the new principles he has learned.

Your effective teaching plans and presentations are often based on a relationship between the parts of a concept or activity and the total picture of the activity. This simply means that before a student can fully understand segments of an activity, you should first provide him with a picture of the whole activity. Once the student sees the whole picture of what is to be learned he can begin to fit the parts into the picture in relationship to each other and to the whole picture as each is explained. As the student begins to understand the parts one by one, a completed picture is provided and the concept or activity is seen in its complete form. For example, consider a student activity for etching a printed circuit board in an electricity-electronics course. It is essential that the directions and presentations complete the whole picture before the etching problem is attempted. The student should understand the total problem before attempting any individual part. One illustration would be that he should understand the concept of a circuit diagram as a part of the complete activity. Without this understanding, the performance of solving the problem may result in failure. In other words, directions and presentations should be made concerning each subtopic so the student will completely understand the total problem of preparing the printed circuit board. The difficulty and relative importance of each subtopic will determine the time and emphasis placed on it.

The whole concept just described and applied is but a subunit or topic of the total content in the student's course in electricity-electronics. Here again, he should have learned and understood this relationship.

Your job as a teacher in this process is one of guiding learning. It is imperative that you always keep in mind that you have a group of students and a knowledge of certain things which these students must understand and be able to perform in order to complete a learning situation. Your responsibility as a teacher is to see that your students acquire this necessary knowledge or mental and manipulative skill.

As you can readily see from the discussion on learning activities and the concept of plans and presentations, you will be required to use a variety of methods, techniques, instructional media, materials, and equipment. It is necessary that you be prepared to adapt procedures to meet changing laboratory situations. You must be aware of the possibilities, advantages, and limitations of each and be prepared to use that which most favorably meets the situation at hand. Teaching methods and instructional media are covered in other Chapters.

Atmosphere for Learning

The atmosphere for learning is so important that you should give it considerable attention. It is not an easy thing to plan, but proper preparation and evaluation during the teaching-learning situation will give you an opportunity to provide for special arrangements. There are a number of principles which should prove quite helpful to you in planning the best climate for learning.

There are actually two aspects to the planning of a desirable learning atmosphere; the physical setting and the emotional quality of the teacher-student relationship. The physical setting of the laboratory learning environment has a great impact on the quality and ease of learning. You should consider the physical setting in respect to providing an atmosphere which makes learning inviting and interesting. A number of factors that have considerable influence on the physical aspects of the laboratory for learning are as follows:

1. Chairs, desk, and tables should be arranged in such a manner that students can see, hear, and be comfortable during your presentation. If a student is not physically comfortable; for example, sitting on a bench for a long presentation, his ability to concentrate on the learning activity is seriously hampered.
2. Sleepy students are slow learners. Temperatures in the laboratory should be such that students are comfortable but the temperature is not conducive to sleepiness.
3. A well decorated laboratory provides an interesting learning environment. Colorful equipment, interesting displays, pictures, and lighting effects enhance the learning climate compared to a drab, uninteresting laboratory. Laboratories can usually be improved when these factors are considered.

4. Your presentation materials are important. They should be planned ahead and all articles or physical necessities for the particular learning activity conveniently arranged. You are not putting on a television "special" as an actor, but you are making similar arrangements so students can easily follow important directions and explanations.

5. Vary your physical settings of instructional media and learning "action" centers. Students who expect a boring demonstration will undoubtedly be bored. If they arrive to find an exciting new learning situation, a positive atmosphere for learning may well have been staged.

The second aspect of the learning atmosphere, that of the emotional relationships between students and teacher, has a similar effect on the learning situation. A student who says, "I can't stand that teacher," is probably referring to this aspect of the learning climate rather than what the teacher knows. The good teacher looks upon his students as individuals. Everything you do as a teacher adds to the students opinion of you as a person and influences the way they react and respond to your efforts to help them. It is not unusual that presentations which are normally clear will not be clear to the student if the atmosphere of cooperation is absent in the laboratory learning situation. Be patient and cooperative with your students and they will most likely respond in the same manner.

Lesson Planning

You have now established a sequence of planning dealing with technical courses in industrial education that lends itself to developing class activities in usable form. This usable form has long been known as a lesson plan. The term LESSON PLANNING means many things to many people; too often a comprehensive written coverage of all the things to be said and done during each class period. In many cases they have been so complicated that they have been restrictive and frustrating on the part of the teacher. At times the lesson plan has been little more than scores of written material taken from textbooks and other resources and read word for word to the class. This is not the meaning of the term being used for discussion here. To better describe the term, consider it to be a "contemporary plan of class activities." LESSON PLANNING WILL REFER TO THE BRIEF WRITTEN ARRANGEMENT OF ACTIVITIES TO TAKE PLACE DURING A CLASS PERIOD ON THE PART OF BOTH THE TEACHER AND STUDENTS.

Perhaps this concept can be better illustrated by shifting the emphasis from the lesson plan to the depth of study and understanding on the part of the teacher. This suggests that the lesson plan become a brief guide for class activities strongly supported by a resourceful and knowledgeable teacher. If you have done your course planning and have developed a wealth of information concerning your technical content, there is little need for you to read to your class. The greater emphasis here should be placed on planning and shifting away from the more traditional concept of a daily lesson. A good industrial education teacher will have prepared a daily guide for class activities from his recent notes on all of the technical material from his course outline. Certainly you will refer to your notes for specifics from time to time, but you should have such a depth of understanding and study of the content that it is not necessary to rely on written material for all types of presentations. It should also be noted that modern lesson planning places more emphasis on instructional media and learning activities than did the traditional approach. This requires a different type of planning, since all of your content is not expressed by lecturing.

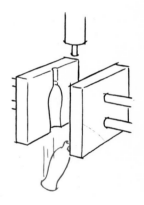

Let's look at a contemporary approach to lesson planning that should provide many ideas and suggestions as you study the development of daily learning activities. If you have your topical diagram, your course outline, and units of study, you are in a position to combine content with presentation. This is where the real planning begins, for you must start making decisions as to how effective each teaching media will be for the various concepts and skills to be presented. These are then organized for your total unit of study. An analysis of a selected sample lesson plan should serve to illustrate the contemporary approach. In this case, a plan for a class period activity dealing with a unit in plastics processing follows:

A. Blow molding processes.
   1. Discuss the main principles of blow molding (notes).
   2. Show sequence slides of the industrial process (slides 1-13 you have prepared).
   3. Discuss plastics materials used in blow molding (pass out list).
   4. Illustrate the internal operation of the equipment (use overhead transparencies 29 through 36).
   5. Demonstrate blow molding to the class (laboratory machine).
   6. Have students cut a blown bottle in half and discuss the terminology related to the product.
   7. Assign outside activity (pass out activity sheet for study of plastic bottles found in the home).

The lesson plan looks simple as well it should, and yet, a further glance reveals a multitude of planning that has gone into its arrangement. It also helps to illustrate the guide lines that have been established for the topic. A lesson plan such as the one illustrated suggests a number of factors that went into its planning and pertain to its use. A number of assumptions should come to light as you review this approach in comparison with the more traditional lesson plan. Let's look at the topics one at a time, which would apply to any technical class:

1. The use of notes in discussion of the topic presumes you have a knowledgeable understanding of the topic which has been outlined in advance. Your notes serve both as a guide for discussion and to keep you on the right track. A revision of your discussion notes should be a continuous process as you review and study the latest in technical literature.

2. The slide sequence presumes you have a narrative to accompany the presentation providing further opportunity for discussion and questioning.

3. A list of materials used in the process is passed out to each student. The opportunity is provided here for students to relate the content they have studied to a further investigation of the specific materials used in the industrial process.

4. Overhead transparencies which you have prepared make it possible to question and discuss the internal operation of the process. This overcomes the problem of trying to picture the process in the student's minds by listening to a lecture.

5. A demonstration of the process on laboratory equipment provides the student with a "hands on" experience as he learns to operate the equipment himself.

6. The laboratory activity in which the students analyze the product again leads to further questioning and investigation. An opportunity to analyze a similar product of the same process provides out-leadings for further study of variations in the process.

7. The use of an activity sheet for an outside assignment allows for freedom of investigation with guide lines for independent study.

You may make a lesson plan for this type of study quite differently, but this is an example of one arrangement which should prove helpful in your study. Of particular importance is the variety of methods and media used for any topic to meet the best learning needs of all students. Perhaps some answers to typical questions will provide you with further insights. Is it easier than the traditional plan in which all the teaching material is spelled out and you follow it directly? No. Does it provide less work in preparing yourself and the teaching-learning material to be used? No. Will it provide the student with a better opportunity to learn? Undoubtedly, since you are searching for all possible media to meet the needs of the modern psychology of learning.

Once all lesson plans have been completed can you relax with a job well done? Certainly not! Inspection and revision according to the latest technical advancements and developments in teaching media require constant changes and up-dating. Is it fun? That invariably depends on how much you love to teach.

Follow Your Planning

The lesson plan is your personally developed outline for teaching. It would probably be difficult for someone else to follow since you developed materials they would probably not possess. Your plan keeps the pertinent materials before you and insures continuity to your presentation. Developing an outline plan should keep you "on the right track," prevent the omission of essential points, provide an interesting learning atmosphere, and avoid the introduction of irrelevant materials. Students have a sense of knowing when you have done your "homework" and appreciate that you have given the same attention to your teaching that you expect them to give to their learning. There is no better road to teaching success than a carefully thought-out lesson planning activity.

Follow your planning. A great deal of time and effort have gone into its development, so use it wisely. It is doubtful that you will ever need to hold your plan in your hand but it should be available to you at all times for quick reference. If your plan gets in your way so your students are more aware of it than the learning situation, it loses its value. On the other hand, a contemporary plan as illustrated would leave little for you to read from or depend upon other than as a guide.

In using your plan as a guide, you should be completely familiar with all the necessary content far beyond that needed for the particular learning situation. Your lesson plan is simply a guide for smooth teaching. Follow it as you would a road map, check yourself as you proceed and revise as necessary when learning conditions require you to detour or even go over the same route twice.

Be flexible enough to adapt your plan to the learning situation. No two classes will be alike, and if you find in presenting the learning activity that your planned procedures are not leading to the desired results, revamp your approach. Be prepared, be thorough, be flexible, follow your planning. BE A GOOD TECHNICAL TEACHER.

Effective Explanations

Another phase of following your plan is that of providing effective explanations. Often this is not an easy task. When planning your explanations, give the most emphasis to those technical directions which are necessary for the student to do the activity or apply the concept presented. It is sometimes necessary to go over your explanations a number of times to be sure the student has a complete mental grasp of what is to be expected.

Good teaching calls for explanations and directions to bring about the desired idea or understanding on the part of the student in preparing him to proceed through an activity. This does not presume that you are just passing along information, but it does indicate that you are presenting explanations that bring about readiness for action. Few students understand all directions or explanations in the same way and if you are to get each student prepared for an activity you must give each one an opportunity to understand and gain an understanding in his own way. Just telling is not enough. You should encourage students to participate freely and openly to help accomplish this goal. Students can help each other. They talk the same language, and sometimes they may convey ideas to each other better than you can. Give the students an opportunity to "get into the act" at every possible opportunity.

Two ways of obtaining student participation are to make your presentations in such a way that they will arouse questions and center learning around student problems. This may be easier said than done. However, teaching experience provides you with insights in gaining student participation. Explanations which often result in a question to complete a concept is one such approach. Another is the problem-centered situation which automatically calls for student involvement, especially when the problem is a student difficulty. Contributions from students from their own experiences will often contribute to solutions which students can understand and appreciate.

Effective explanations take skill and practice in the art of communication. Unless you take advantage of student involvement and questioning, the communication will all be one sided. How many times when you have been reading some technical literature have you wished someone was around whom you could ask a question or discuss a topic you did not understand? Perhaps your study was interrupted or halted. Effective teaching provides a smooth continuity of learning. Your presentation may be the answer.

Let the Student Know Where He is Going

Whatever type of instruction you are doing, part of your planning for presentations should involve guide lines for the student to view the course as a whole and the place of each unit in the course. If your planning has been done effectively, the students will know where they are headed and what to expect next. Confusion on the part of the student as to where he is going hampers the learning situation. Be explicit in your planning to give him direction.

One of the most satisfactory methods to accomplish this is to provide the students with a modified course outline drawn from your extensive outline. This would probably include the units of study in sequential order as you expect to present them. It should also include such items as required laboratory activities, reading and study resources, textbook assignments, and evaluation information. The students are more likely to grasp the total idea of the course when they are presented with a comprehensive preview. Not knowing what to expect from unit to unit or from day to day deprives the students of the opportunity to link principles and activities together for a total understanding of the technical course. To understand a course in power, for example, would indicate that the many concepts have been brought together as a whole. Fragmented, unassociated principles do not lead to this type of understanding.

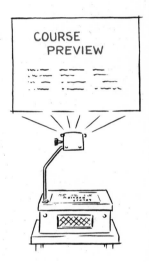

COURSE PREVIEW

Plan your trip to "learningville," share it with your students, explain stop-overs and side excursions, discuss who shares responsibility for reading the map, emphasize the need for cooperation, and you will have a successful trip.

Follow Your Presentations with Activities

Considerable attention has been given to the planning of learning activities. Now consider the place of such activities in your plans and presentations. There is probably no greater learning activity than the "hands-on" experience of going through a process, performing an operation, or participating in an experiment. Yet the value of your learning activities lies in their use at the correct time and place in your units of study. Preparing a mold and pouring a casting proves of little value to the student when it is out of context. In other words, it is little more than a rote manipulative activity (routine activity carried out mechanically without understanding) from which you could expect little learning to result. However, if the same activity took place in a sequential order as indicated in the illustration of a suggested lesson plan, the result in learning should prove to be quite meaningful.

In your planning of presentations it is a good idea to give considerable thought to following your presentations with student activities that will reinforce the concepts or principles to be learned. The type of activity may be as variable as the content itself. The important thing to remember is that a student activity is actually a follow-up to the presentation or explanation of an idea, concept, principle, or process discussion. It is one of the most meaningful challenges to you as an industrial education teacher to devise student activities following presentations that will bring out the ultimate in understanding and comprehension of the topic involved.

Discussions and Questioning

Although further treatment of discussion as a method of teaching is presented in the Chapter on Instructional Methodology, it is well worth consideration at this point since discussions and questioning relate to teaching presentations. Discussion is one of the most effective means of student participation in the process of presenting and explaining. It is the sharing or interchange of ideas among all members of the class. As a teacher of technical subjects, discussion assists you in two major ways. First, after you have given a presentation, it is a valuable aid to you in locating and clarifying any misconceptions on the part of class members. Too often students do not realize that they have misunderstood a particular principle, and therefore, will not ask questions. A discussion of the topic may well reveal this misunderstanding to the student and to you as well. Clarity and insights may then be obtained.

Second, the discussion technique used during presentations may help you clarify specific directions that the student has not understood. It is one way to help overcome the problem so often

encountered when you observe students using equipment improperly or engaging in an activity with little understanding. Discussions give students an opportunity to relate to each other on the same level, giving the slower student a chance to see where he has failed. A few suggestions may help you in evaluating the use of discussions and questioning in making presentations:

"Be the leader in conducting discussions." Your role is that of making discussions of value to all students and that the results clarify for each student what he is to learn or perform. It is your responsibility to see that the discussion is orderly and meaningful, not just a "talk session."

"Provide for total class participation." During the discussion of technical topics one of your main tasks will be to encourage, question, and stimulate each student to make his individual contribution to the group. Some students shy away from participating in discussions. Get them involved by asking for their opinion or explanation.

"Make discussions part of your teaching plan." Any discussions should move the class closer to the goals to be achieved. You must make plans for discussion topics and see to it that no individual monopolizes the period. Too often two or three students do all the talking. You can direct questions to other students that will bring the rest of the class into the learning situation. Part of your responsibility in planning discussions is to keep the students on the "right track." If the discussion varies from the topic, bring in new ideas and ways of solving problems that will guide them back in the right direction.

"Observe your students carefully." Observation during the process of discussions is of critical importance. You can often determine expressions of agreement or misunderstanding from student expressions. Take the opportunity to ask why a certain student agrees with a concept or why another does not seem to understand a principle. Let them question themselves and encourage them to relate to the class what they are getting out of

the discussion. Like any other means of helping students get a clear understanding of directions and explanations, the success of the discussion is measured by how well the student is able to make use of the results of it in technical performance, analyzing, and evaluation.

Discussions can be overdone. You must evaluate the progress critically and determine just when the time has come for students to put the "understandings" into action in a laboratory activity.

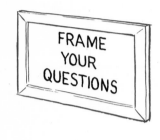

The art of questioning is one of the most effective teaching devices in providing for clear, understandable explanations. Using the Socratic teaching technique of asking questions but seldom giving direct answers, causes the student to rely upon himself to gain understandings. Here, the use of questions makes the student analyze and study as you lead him into byways from which he is able to learn his own way back. It provides a mode for thinking that is left void if the student is given all the answers. The process of directing student thinking by questioning may generate the curiosity and interest necessary for good learning. Questions cause the student to put ideas together into complete patterns and arouse his curiosity and desire to explore the problem further.

The teacher who is able to develop the ability to question skillfully is usually a good teacher. It takes time and practice to become skillfull in being able to ask the right question of the right student at the right time. As you develop this ability you find that it is a two-way street. You ask questions as a teaching device in making presentations. Your students ask questions because of doubt, curiousity, or lack of understanding.

It is important that you provide a laboratory learning atmosphere in which students feel free to ask questions. You will find that student questions may be more effective than your own for it may be difficult to put yourself in their level of understanding. Their questions may point out places in your directions and explanations that need improvement. Student questions usually indicate perplexity and doubt that requires further explanation. They may even indicate the nature of the student's difficulty.

You must also be able to quickly analyze the nature of students questions and think ahead to provide for an effective follow-up. If trivial questions arise, dispose of them as courteously as possible for they are distracting to the topic at hand. Many students are reluctant to ask questions which may expose their ignorance or misunderstanding. Make it as easy as possible for the less capable students to enter into discussions with questions and let the class know the value of the questions asked.

Let your students contribute, since their confidence is buoyed (kept afloat or sustained) by an interchange of ideas and questions, making them feel they have played a valuable part in the learning process.

Avoid Confusion - Make it Clear

The industrial education teacher has an extra responsibility in making presentations because of the very nature of the complex technical concepts with which he often deals. It is of extreme importance that you make your explanations clear and avoid confusion on the part of the student. Your presentations are the guides by which the student takes the shortest route to understanding and effective performance. When your explanations are well given, they become a clear, definite line of action for the student to take in performing an operation, experiment, or any of a multitude of activities. When poorly given, they lead to confusion and wasted time.

Be precise. The process of explaining is the process of clearly conveying technical concepts or competencies. It is the process of "showing and telling" the student in such a manner that a minimum of confusion exists when he applies the principles he has learned. Provide the student with clear explanations, for he looks to you for leadership for the why and how of doing things.

If the student is engaged in an activity, say milling a gear, more explanations may be needed as he proceeds to guide him to successful completion of the operation. This is just one of the many parts of the whole learning process, although treated separately, that you as a technical teacher must relate to the learner.

The experienced teacher has learned how to convey the action part of learning by the use of concise explanations. If you want someone to learn something, tell him exactly what it is you want him to learn and he is more likely to learn it. Be precise and direct in your explanations. Long, drawn out theories that do not apply to action on the part of the student delay the learning process.

Explanations should include both the "why" and "how" of the activity. A student may well initiate the doing aspect of a learning situation, but when it is explained why he should do it, both interest and attention to the activity will probably be upgraded. Be sure you give him every opportunity to understand and perform at his best potential by making your presentations clear.

Helping Students Study

The industrial education laboratory provides one of the finest situations for teachers to assist students in learning how to develop good study habits. In many other areas of the educational program the student must do most of his studying outside of the classroom or at home with little or no supervision. The laboratory situation presents a number of opportunities for you to establish supervised student study.

Many of the activities and assignments you plan should call for individual or group learning sessions in which the student is required to use a variety of study techniques to solve the problem or complete the assignment. The learning activity described earlier on wood identification is a good example to follow for supervised group study. In such a situation, directions and explanations have been given and the group works together on the assignment under the teacher's supervision. This is not a time when the teacher grades papers or takes a convenient rest period. It means just what it says, supervised study, and requires considerable work on the part of the teacher. You would observe the class as they prepare the specimens and analyze the wood structure, making comparisons with the data on their key for identification sheets. The students will soon learn that they can look to you for direction when they meet situations they do not understand. You are their resource person. As they begin to develop an ability to make an analysis by themselves they can proceed independently.

Supervised study requires a definite plan for the student to follow with your assistance in helping him over any hurdles as he proceeds. This applies to group or individual study. While studying under your supervision, you are able to see the difficulties he encounters and make suggestions, review, demonstrate, and provide a better framework for his study habits.

The same applies to assignments for study outside of the laboratory. Assignments should be designed so that each student is given direction for study. It is your responsibility to help the students get a clear idea of what the assignment will do for them. You should assist the student in getting the most out of his reading, help him learn how to follow directions, look up resource material, use the library, and give him directions and examples of how to study.

Review Questions - Chapter 6

1. Give a number of examples of how a teacher may establish a good atmosphere for learning in the laboratory.
2. Explain the advantages of supervised study. Why is supervised study so important to the student?
3. Why are ideas alone of little value in the learning process?
4. What is meant by guiding the developmental learning process that takes place in the student?
5. What are the two main aspects of planning a desirable climate for effective student learning?
6. What qualities are found in a good teacher that reflect in an eagerness to learn on the part of the student?
7. Explain some disadvantages, as you see them, with the traditional preparation and use of the lesson plan.
8. Why is it imperative that you have all of your teaching materials ready and in order when you plan to make a presentation?
9. What reasons can you give which illustrate that the preparation of a contemporary lesson planning activity requires more work than the traditional lesson plan?
10. How have the rapid changes in our technology caused the teacher to make changes in his lesson planning and presentations?
11. What are the advantages of being flexible enough to adapt your planning to specific learning situations?
12. During a presentation of a technical topic, why and in what ways can you obtain student participation?
13. How can you provide the student with insights and information as to the direction and goals for learning in a technical course?
14. What are the important aspects of following presentations with learning activities?
15. What are the two main ways in which the use of discussions may assist you as a technical teacher?
16. Explain what is meant by the Socratic method of questioning.
17. In what ways might you encourage questioning on the part of students who may feel too embarrassed to ask?

Suggested Activities

1. Plan a learning activity in your technical area of interest that is designed to place heavy emphasis on assisting students to learn how to study effectively.
2. Develop a contemporary lesson plan for a selected technical topic which includes the use of some instructional media, teacher presentation, student participation, and an out-of-school assignment.
3. Make a listing of topics from a unit in your technical area and suggest follow-up student activities which you feel will best reinforce the concepts to be learned from each topic.
4. Write a short paper on the use of discussions and questioning for a technical topic. Use as many references as possible to support your ideas.
5. Select an article from a professional technical publication which depicts a new process or technique in your area of interest and make an outline for a presentation for a specific level. Be sure to interpret the content of the topic for the level of students to whom it would be presented.
6. Search through a number of old publications or ask your instructor to help you locate a traditional lesson plan that included everything that was to be said and done during the class period. Indicate the ways in which that lesson plan would not serve today's needs.

Chapter 7
STUDENT MOTIVATION

Think back over the years and try to remember specific occasions when you had a "burning" desire to learn something. It need not be in a classroom or even in a school situation. Are you able to pinpoint any of the reasons why you wanted to learn something so much? The main idea for such a question is to give you a chance to think over the "why." Perhaps it was to perfect a hobby, become more skilled in a particular sport, to meet the requirements for a job, or perhaps an unusual interest in a school subject. A question of this nature should provide some indication of why you were motivated to investigate, inquire, or feel a need to further enlighten yourself about something.

Motivation is a "slippery" and elusive term, and yet, this is what it is all about. Do you recall a particular class or course in school in which the time always seemed to pass too quickly? You may also recall some other classes in which you could hardly wait for the period to end. Undoubtedly one of the prime factors differentiating the two situations was "motivation." It is said to be elusive because it is certainly not something that can be factually pinned down and effectively put to use any time you desire. Student motivation must be studied and placed on your conscious awareness list during all teaching situations. A student who is not motivated is bored, and a student who is bored is likely to learn very little. Hence, motivation becomes a very important topic for consideration on the part of the industrial education teacher.

The industrial education teacher is often in an enviable position when it comes to student motivation. A main reason for this position is that the technical content of his courses is usually exciting to the students. So much of the technical content involves acting - thinking and doing - that it is unusually appealing by its own nature. However, this is not enough itself. As a teacher, you must be continuously alert to using procedures which will make your students want to learn. Without a desire to learn, there is little or no learning. Therefore, you must actively engage in bringing about a desire to learn by arousing the interest of your students. Once the desire to learn has become enhanced in the student, learning usually proceeds under the student's own power without the necessity of being constantly prodded by pressure or fear of failure. In planning presentations and activities, motivation devices should be "built in." A learning situation so planned creates a readiness, an eagerness, a desire to learn which is not satisfied until learning is achieved.

Enthusiastic Teacher - Interested Student

It is difficult to hold students "spellbound" during a lecture or presentation. However, a good teacher has many devices, incentives, methods by which he can captivate his students during the learning situation. One of the most outstanding of these is that of being highly enthusiastic about your technical subject matter. Enthusiasm rubs off; don't hide it from your students. For example, most students gain an excitement for learning when you show a terrific excitement about teaching and the content. Technical drawing may provide a good illustration for this point. It can be very interesting and exciting, or it can be a very dull course. Certainly much of this hinges on the content, the activities, and class organization, but a great deal also hinges on the enthusiasm displayed by the teacher. Although both go together in the teaching-learning situation, concentrate on the latter for purposes of discussion to see how enthusiasm may provide motivation. Some suggestions for your technical drawing course are as follows:

1. Be a moving teacher who is always active, whether you are making a presentation about a problem in drawing or supervising student study. If you sit at your desk all of the period, students may not be motivated enough to even notice you are there.

2. Put some humor into your presentations and in the use of your instructional media. Certainly this can be overdone, but the right amount adds to the interest and motivation of students to be alert and enjoy your teaching. It is hard for a student to fall asleep at his drawing table when he is having fun. Did you hear anyone say learning should not be enjoyable?

3. Present news of the day. Start off each drawing class with something new and exciting from the industry or society related to the course. Your students will begin to look forward to each class with a better attitude and degree of motivation to learn.

4. Try a little friendliness. Enthusiasm and friendliness get along quite well together. When your drawing students realize that your enthusiasm is related just as much to them as individuals as it is to the course content, they may well return your interest in them by taking on a greater desire to learn. An enthusiastic pat on the back for a job well done or a concept understood goes a long way in maintaining an interest in learning.

Do as much as you can to also show your enthusiasm through your laboratory activities. Students can tell if you have thrown yourself enthusiastically into the preparation of your course. If your activities and presentations are the same old thing year after year, the word gets around. It also gets around if your course has been an exciting challenge to students because you have made your necessary revisions and kept your content contemporary in nature. Some of your students may show a high degree of enthusiasm and interest. You should not allow this to diminish. Other students may show boredom and be antagonizing which you must try to convert to enthusiasm and cooperation. The attitude of the majority of your students, however, will not take on either of these extremes. You must assume the responsibility of convincing them the content is important and worthwhile. Your enthusiasm will go a long way in helping them become interested and eager to learn.

Making Learning Exciting

The industrial education laboratory poses an atmosphere of excitement and interest for your students when it has been planned on a contemporary format. All of the tools, equipment, machinery, testing apparatus, or variety of supplies set a beautiful stage for exciting learning experiences. The inquisitiveness on the part of the student to learn about the processes or concepts he may identify within your laboratory is an ideal starting point. Make use of your presentations and activities to whet this interest and curiosity.

One of the best means by which you can make learning exciting is through the use of a wide range of instructional media and student centered activities. A time to lecture and a time to work no longer satisfactorily meets the needs of the technical courses offered in our schools today. Things are too complex and learning media are more available than ever before, to use such a shallow approach. Exciting learning is based upon the philosophy of meeting the needs of the learner and placing the responsibility for learning in his hands for action and performance. When the student is introduced to a new topic or concept through a variety of techniques and sees the immediate gains through the planned activities, it is usually quite easy for him to become excited about learning. Again, an illustration may serve to make the point.

Suppose you were preparing a learning activity on precision grinding in a metals course. You could make it quite dull or very exciting to the students. The dull approach would be a lengthy lecture and a lengthy demonstration on a surface grinder. To make the same topic exciting, you would probably establish an immediate objective, such as sharpening a particular milling cutter that the students are using in the laboratory. Your next step would be to plan a total presentation which would involve the students as much as possible. For example:

1. A colored slide presentation of microscopic views of dull and sharpened milling cutters and other ground surfaces.

2. A motion picture on precision grinding of cutting tools.

3. Use the class to make a television tape of the setup and grinding operation.

4. A student presentation of a chart he has developed depicting the shapes of wheels used for grinding cutters.

5. The use of overhead transparencies to show the use of jigs and fixtures required for grinding cutters.

6. A student activity sheet that calls for the student to make sample ground surfaces by using the various grinding techniques.

7. Set up a group situation in which the class is responsible for demonstrating the grinding of a milling cutter.

8. Outline a research report for each student on some aspect of the many types of precision grinding.

This illustration would not be desirable for every teacher, but it brings out some of the many ways in which your planning may motivate student learning and make it exciting. The main points to remember to make learning exciting are the use of a variety of modern teaching media, considerable student participation, and an immediate goal in the student's mind.

Students Should Recognize Goals

How many times have you heard a student ask, "why do I have to study this?" and a response too often given by a teacher, "because someday you will need it." This type of situation should not exist in the first place, because student recognition of the goals established for units and the total course should be part of the developmental learning process. If the plans and presentations for the introduction of the course include immediate gains and long range goals, the student is in a much better position to know where he is going and why.

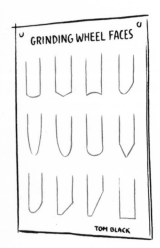

Suppose a golfer is heading for the first tee of an 18-hole course. Why is he there? He is probably there because he enjoys playing the game. Depending upon his experience, he probably has an immediate goal of trying to make par for the course and perhaps some long range goals in the improvement of all aspects of his play. If he is serious about his game, he will consciously be trying to improve his putting, overcome that nasty slice, or become more consistent with his chip shots. He recognizes where is is going, knows what it takes to get there, and is certainly aware of the rewards for practice, study and attention to his game; satisfaction and pride in his efforts.

How similar to the game of learning, especially in industrial education where mental and manipulative skills are so important. Yet the reasons for learning, the immediate gains and long range goals, are often overlooked. Be sure you provide your students with planned opportunities to discuss and understand the purpose for their study and learning activities. An example might be a particular activity where a student learns how to use a pyrometer in measuring the temperature of molten metal. This immediate goal is of such a nature that it can be obtained within a short period and takes the student a step nearer his long range goal. In this case the long range goal might be a particular objective, such as learning the ideal pouring temperatures of various metals in making castings. This long range goal leads to a course objective, which in this case might be to understand the concepts and skills necessary to perform efficiently in the foundry industry. The emphasis on these goals may vary accordingly in a technical, vocational, or industrial arts course, yet the principle remains. The student must see the relation of each short range goal to his final long range and recognize the progress he is making.

Students Experiences are Meaningful

Every experience your students go through will have a definite effect on their attitudes and understandings. It is especially important that these experiences are always made meaningful. "Busy work" is a waste of time. Meaningful experiences on the part of the student are highly motivating and lead him from one activity to another with enthusiasm.

The responsibility you have as an industrial education teacher in this respect is to plan your learning activities in such a way that they are relevant to the learner. When he sees the meaning behind the performance of the activity, you have the learning experience well under way. The problem of making experiences meaningful involves a number of factors related to motivation. If you have done your job in course planning and presentations, and selected activities with the student's needs and goals in mind, the pleasant satisfaction of having gone through a variety of experiences will make them meaningful to the student. When he understands why they are important and receives satisfaction from their accomplishment, the meaning behind the experiences becomes part of the student's attitude.

One good experience leads to another. When students see a process, a piece of equipment, or an operation they have seen before, they may tend to become interested in that part of the learning activity. The student's background gives you some indication of what he is familiar with. Take the opportunity to make use of this familiarity when it can be seen that it may add confidence in his ability to perform an activity. Your students have a wide variety of experiences of which you can take advantage. They can provide you with quite a resource of information about themselves. Give them an opportunity to talk about their experiences. If a certain activity reminds them of an experience they have had, encourage them to express it. The analogies they use may prove helpful to other students. If a student says, "A giant molecule reminds me of a tinker toy," let him explain it further. As more and more student experiences provide good illustrations, their interest in meaningful experiences continue to increase as they relate to familiar and pleasant activities they have gone through previously.

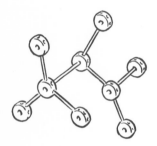

It also provides you with a good opportunity to locate concepts and processes that they have already experienced. Many times you are then able to keep from repeating presentations or activities which they already understand, which might cause them to lose interest. EXPERIENCES ARE MEANINGFUL AS THEY LEAD THE STUDENT FROM WHAT HE ALREADY PERCEIVES TO NEW CONCEPTS. If you are able to provide such meaningful experiences, you have gone a long way in motivating student desire for learning.

Success is Highly Desirable

Every effort should be made in technical teaching situations to anticipate where failure may occur in any phase of student learning. A student who continually fails in making satisfactory progress in his study, on tests or examinations, on laboratory activities, or any other part of his pursuit of learning, may soon become disappointed and uninterested. He will become less and less motivated to try harder and put forth effort to learn. Since all of your students will have different abilities, both mental and manipulative, it is imperative that you give them every opportunity to succeed. Any technical course you may teach will be loaded with obstacles in the form of learning situations which each student will attempt to meet with some degree of success or failure. Every student is going to go right on living whether he has success in your course or not. However, you may have a considerable influence on his life due to your interest in encouraging him to be successful in his course work. In other words, a little extra effort on your part to assist in his learning may pay off great rewards on the part of the student to try, and be more successful in other situations.

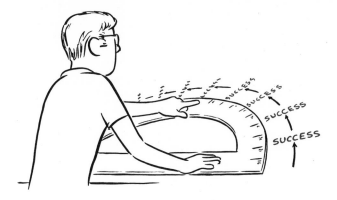

Isn't that what learning is really all about? It is not a matter of stretching a wire across the valley between education and life where you, as the teacher, watch each student attempt the walk. Then as each falls off somewhere along the line you record a grade for his effort and distance obtained. It is really a matter of establishing good laboratory learning situations in which every student who participates has an opportunity for success in many possible forms. This is where success and motivation become close relatives. Even your poorest student may be motivated to accomplish tasks he had never thought were possible if you encourage him and give him that extra time required. Let your students know that you have confidence in them and that you are there for the purpose of helping them succeed in learning. They will respect you for it and in many cases try harder for fear of letting you down.

Success is a matter of degree. Failure is a matter of losing. Motivation eliminates losing and establishes a student's own guide lines for learning. They will not be the same for all students, but a little success is far better than all failure. Your part in the motivation of students is fairly well outlined. Some will require little or no encouragement to learn while others may require continuous motivation through the many possibilities discussed. However, another facet of motivation and learning is the willingness of the student to cooperate in laboratory activities.

Class Control and Discipline

Motivation is a close relative to both class control and discipline. If all students were highly motivated to learn, there would be few if any problems of class control. Many of the factors dealing with motivation of the student to become excited about learning are in direct proportion to the problems of class control and discipline. However, there are other factors also involved which are part of the teacher-student relationship.

The industrial education laboratory provides an ideal atmosphere conducive to good learning and fewer discipline problems than most other areas in the educational curriculum. In a good technical teaching-learning situation there are too many "active" things to be done, too many interesting things to be investigated, and too much to become excited about for the student to "fuss around" or "get into trouble." On the other hand, you will not always have an ideal teaching-learning situation and you will undoubtedly have problems with students from time to time. Student problems of an individual nature will probably require private conferences in which you may be able to locate the problem and assist the student in making changes in his behavior. In other situations, it may be necessary to refer the student to the school conselor if the problem is of such a nature that you are unable to resolve the necessary behavior changes.

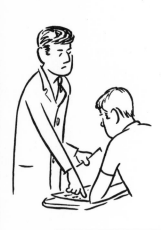

Problems of class control are of a similar nature. When you are aware of situations that have developed within the class that are not conducive to overall learning, take the time to discuss such problems directly with the class. Always place the emphasis of class cooperation on a positive note in such discussions. You are working with students not machines, and you must get to know them well to secure their cooperation. In any cases where mental or physical handicaps disrupt normal class procedures, you should call upon a counselor or specialist who is in a better position to work with students to help solve such problems.

The major emphasis of class control should be placed on the positive attitudes and actions you expect from your students. Discipline often refers to a negative reaction in which a student must be punished for something he has done wrong. There is no place for punishment in the industrial education laboratory learning situation on the part of the teacher. If a problem of such drastic measures did arise, it should be taken out of the laboratory situation and placed in the hands of the administration or counselors who are in a much better position to cope with such problems.

Perhaps a few suggestions that deal with class control will prove helpful. You will notice that discipline in any of these suggestions is a positive and firm gesture of student responsibility.

Encourage all students in class participation. When they are actively engaged in activities or discussions, you are developing real interest and intent on learning.

Insist upon the rights of each individual in the class and his respect for the rights of others.

Never criticize or be sarcastic to students in a personal sense. When criticism is necessary, make it a private discussion. Use a straight forward correction of student mistakes in an impersonal manner. Be courteous to your students.

Be decisive in your explanations. Once a decision is made, follow through and give consideration to all factors involved as you act with conviction. Your students will respect you for making fair decisions and abiding by them.

Be fair to all students and show no favoritism. Never insult the whole class by the wrong doings of a few. Treat each case individually.

Insist on cooperation among members of the class and show evidence of your cooperation in their laboratory activities.

Provide rules and regulations for class conduct which the students understand and have been given an opportunity to participate in the formation of such regulations.

Be careful of penalizing students for making mistakes. This is part of the learning process and negative discipline will usually cause students to hide errors or mistakes from you.

Show a personal interest in each student. Get to know their interests and problems and show how delighted you are with their achievements. Interest in student performance usually goes along with disciplinary problems.

Motivated students will seldom cause difficulties in the learning situation. Take advantage of all your resources to keep learning activities moving and interesting. Have something challenging for all students to do at all times. An idle student will usually think of something to do if you have not kept him active in learning, and what he thinks of doing may cause disciplinary problems.

Control your class by motivating students to want to learn and take on a desire to be loyal to you as a teacher and friend. It is harder for a student to hurt a teacher by "stepping out of line" when that student knows you have a genuine regard and friendliness toward him.

Learning Should be Enjoyable

What fun it is to have a deeply enjoyable learning experience. What fun it is to study the intriguing technological developments of modern industry, to operate machines with precision, to create new products, to learn a new technical skill, to understand a complex scientific concept. To list them all would fill a book. Yet, learning should be enjoyable.

We sometimes hear students say, "I hate to go to school," or "I certainly don't like that class." Why? Possibly because the particular subject matter is not appealing to them. Perhaps because the climate of the learning situation is dull. Maybe because they were not able to relate the content to their own lives. Whatever the reasons may be, learning should not be boring and dull. It should be enlightening and fun. A great deal has to be learned about the enjoyment and motivation for learning, and yet there are many known factors that can easily be applied. Many have already been discussed, but add a few more to the list which may prove helpful to you as an industrial education teacher.

A happy and smiling teacher has quite an effect on his students. Happiness rubs off. It can become part of the expected climate for learning and keep the students in a good mood. A few funny stories or a joke now and then may add a tremendous incentive to learn, just because the students like to be around you and hear what you have to say. Of course this can be overdone, but look back at many of your classes and see if it wasn't underdone. Education is no joke, but a lack of desire to learn is no joke either.

Do everything you possibly can to relate your course content to the life of the student. If the content is not meaningful, it should not be there. If it is, it certainly has some relation and meaning to the students who are in your class. Provide for the relationship by giving examples and reaching into your own experiences for meaningful opportunities you have had to learn. Why do so many people sign up for ski schools, bridge lessons, or swimming sessions? Have you ever been in such a class or observed one in operation? Why are the students enjoying learning so much? No need to answer, for you already know. They have their set goals, they have a desire to learn, they relate themselves to the content, and they know where they are going. Try to establish the same type of desire and motivation in your class. Learning how to use an ohmmeter, operate an extruder, or develop a series of offset plates can be as much fun and rewarding as learning how to ski, play bridge, or swim. It must be the atmosphere. It is fun to study when the conditions for learning are set in a motivating environment.

You can help make learning your technical content enjoyable by:

1. Discussing and illustrating the meaning of the content with your class.
2. Giving positive reactions to student questions and problems.
3. Using many media when you make presentations.
4. Providing the latest in resource material, equipment, and supplies for student study.
5. Being cooperative, fair, and friendly while working with students.
6. Relating all aspects of the course to the needs and interests of your students.

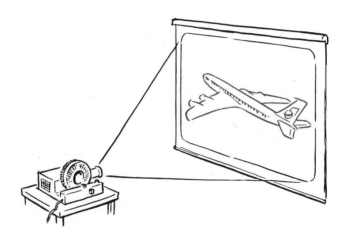

Review Questions - Chapter 7

1. Why can it be said that motivation is an elusive term?
2. What factors would be quickly noticeable if you visited a technical laboratory learning situation in which the teacher displayed a high degree of enthusiasm?
3. Why is an industrial education teacher usually in an enviable position concerning student motivation?
4. What effect does "desire" have on learning?
5. How does an atmosphere of excitement in a technical laboratory help promote student motivation for learning?
6. Why is the concept of "a time to lecture and a time to work" no longer satisfactory for the best learning situations in contemporary industrial education laboratories?
7. Explain how student involvement in learning presentations

and activities serves as a motivating device.

8. Explain why student recognition of the goals or purposes of learning activities is necessary for responsible learning.

9. Why are immediate gains and long range goals often overlooked when teachers plan learning activities?

10. Why should learning be enjoyable rather than just plain serious business?

11. What factors would you contribute to student failure in learning activities which may be harmful in his long range ability to learn?

12. Explain why meaningful experiences on the part of the student provide for a greater ease in learning.

13. In what ways can you make technical course content enjoyable to the student?

14. Why is it more difficult for a student to get into trouble when a teacher shows extreme friendliness and loyalty for the student?

15. What should you do when students with mental or physical handicaps cause difficulty in a laboratory learning situation?

16. What can you do as a teacher to bring about and continue good class control?

Suggested Activities

1. Prepare a technical laboratory activity for your particular specialization which includes concepts that will assist in student motivation.

2. Visit a local school and discuss the problems of class control and discipline with an industrial education teacher. Prepare a list of the suggestions he has for class control and compare these with the principles you have studied.

3. Visit with an administrator or counselor at a local school and inquire as to the responsibilities expected of an industrial education teacher concerning discipline. Ask how the counselor and industrial education teacher cooperate on disciplinary problems.

4. Design a teaching presentation for your technical course for the sole purpose of stimulating interest and excitement about learning some industrial concept or principle.

5. Write a short paper, using the library and reference materials, on the topic, "Student recognition of goals for learning." Use illustrations that deal with learning in the industrial education laboratory.

Chapter 8
INSTRUCTIONAL METHODOLOGY

It has long been recognized and stressed in industrial education classes that a person should use the proper tools for each application. It has further been emphasized that a good understanding and a degree of skill in using tools and equipment is necessary for outstanding performance. A technician knows what equipment is necessary and what tools to select for a particular purpose. The industrial education teacher is in a similar position. He has a variety of teaching methods at his disposal and, if he is an effective teacher, he knows which ones to use for a particular situation.

The analogy may appear simple, yet the concept of meaningful instructional methodology is quite complex. It is not a matter of today you show and tomorrow you tell. Teaching methods require study, patience, and practice to become a part of the personality and conscious being of the teacher. To tighten a threaded fitting, a plumber selects a suitable wrench. To bring about a particular behavior change on the part of a student, a teacher needs a complete set of available teaching tools. If students were as standardized as threaded fittings, the selection of teaching methods would require but little study. Since students are all different in their attitudes and their abilities to learn, an appropriate method of teaching a concept for one may be inadequate for another.

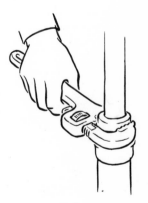

Instructional methodology refers to the ways and means a teacher adopts to guide his students through the various parts of learning activities to the accomplishment of desired goals. In reality, it is actually little more than the means by which the teacher assists the student in solving problems. Once your students have obtained this mental and manipulative ability to solve problems, they are on their way to satisfactorily meet the needs of most any learning situation throughout life. How else can you define the combined efforts of teaching and learning? To learn how to do something by memorization or practice is one thing, to learn how to learn while participating in learning activities is quite another. The first is somewhat dead-ended since it does not necessarily lead to an ability to tackle further problems. The latter is open-ended since it provides for learning something now and also prepares the student to learn with ease outside of a formal laboratory situation. For the sake of providing the best teaching methods possible, concentrate on the second situation.

Problem Solving - The Framework for Methodology

The term problem solving has been used over the years to mean many things. For your purposes of studying and planning teaching methods, it should become the framework in which the many methods of teaching are related to the goals of student learning. In other words, your concern as a teacher should be directed toward both the selection of good teaching methods for any situation along with the development of problem solving techniques as an overall method of assisting student learning.

Think through the process and see if you can apply an analysis of the problem solving method to industrial education laboratory teaching. You can then have an opportunity to delve into the many individual methods in depth. It is, first of all, a series of steps through which the teacher may lead his students in the study of the subject matter of a particular course. It means that topics or content areas would be studied as problems rather than as assignments from the teacher or chapters from the textbook. The problems should be developed through felt needs and difficulties of the student as well as through the value placed upon subject matter. Thus, the subject would not be taught directly, but the teacher would use questions to guide the student's thinking through the steps of the problem solving sequence. Secondly, it is the pattern the student should use in developing his own work; in preparing papers, planning operations dealing with machines and materials, and experimental investigation. The student should be guided by the teacher in formulating and defining his problem, in discovering what data he needs, in arranging material and developing procedures, and in providing means for constructing or testing his ideas.

Your first step in teaching through problem solving or reflective thinking deals with the problematic situation. It must be remembered that problems do not arise by themselves, that your students have few if any problems when the course begins. Problems arise from the process of interaction of the individual with his environment, both physical and human. In the laboratory, the materials, resources, tools, machines, equipment, students, and teacher become the environment, and each student will operate and develop his problems in this environment. The teacher should serve as an intermediary to help the student bridge the gap between his own experiences and those skills, attitudes, habits, and knowledge that society expects the young to acquire. The difficulty for the student in bridging this gap is that he does not know exactly what you or society expects of him, nor does he have enough information and experience to clearly define his own problems and goals. As a teacher, you are in a much better position to understand the expectations of an industrial society and thus assist the student in locating his inadequacies and help resolve them.

The student and teacher should work together, if the problem solving method is being used, to develop a plan to locate the student's deficiencies and provide experiences to meet the proposed objectives. It should be remembered that the problems or experiences that the student is going to undertake be cooperatively planned. Difficulty can often arise when experiences are planned completely by the teacher before the class meets or completely by the student after the class is underway. With these thoughts in mind concerning the formulation of the problem solving method, look at a practical illustration.

During the progress of a course in electricity a student questions his teacher concerning the possibility of developing a doorbell system that would ring in the basement and garage as well as in the living area. The student had picked up the idea as a result of his family's concern over not hearing the doorbell while in the basement and garage, and from a recent class presentation on electrical circuits.

The teacher had an opportunity to cooperatively develop a problem solving situation and took advantage of it. He questioned the student further to see if they had a mutual understanding of the problem and if the objective was worth pursuing.

Since the teacher had observed that the problem was earnestly instilled in the student's mind, he took up his part by suggesting that the student look for factual information that would be helpful in books on house wiring. He suggested also that the student design and construct a model of a three-station doorbell, incorporating modern materials. Stumbling blocks which were met during the activity were resolved.

The teacher suggested that the student report to his class, and demonstrate how a three-station doorbell system would be wired and how it would operate. He also asked for a written evaluation of the activity.

The principles and illustrations just discussed should provide you with an understanding of how problem solving becomes the framework for teaching methodology. It should be the climate or framework from which all other methods used in teaching are drawn. In other words, problem solving is not an isolated method in itself, but rather an overall method which includes the various methods generally classified as explaining and telling, discussion, demonstrating or showing, and performance. The illustration presented a situation in which all of the methods of teaching were actually used; before, during, and after the individual student activity. It must be understood that every learning situation involves a problem to be solved. Without a problem there is nothing to be learned. Of main importance is the selection of teaching methods to assist the student in solving any problem, whether it is initiated by the teacher from his presentations or planned student activities, or a problem that the student identifies. Just what methods do you have at your command to meet the needs of learning; of assisting students in solving problems in their development toward an objective?

Your Resources in Methods

Industrial education courses are somewhat unique in the wealth of theoretical and technical content and the range of equipment from simple tools to highly complex mechanisms. You must use all the resources at your command to make your instruction meaningful and vital for your students in order to meet the demands of mental and manipulative development. One of the best characteristics of a good teacher is that he knows his resources and uses a wide variety of methods and instructional material. He selects the method or teaching technique which best accomplishes a particular goal or meets an existing situation. As you study the many possible teaching methods, you should direct your attention to these resources as a total package of instructional methodology from which you may withdraw in any desired form the resources needed for any situation. Although they are discussed individually, it should be remembered that in most teaching situations many methods will be used in combination and often at the same time. You have many resources in methods of teaching and you must keep in mind that they are means for meeting the needs of your students in directing their learning toward the desired objective.

It is also important to remember that there are no sharp lines of demarcation between methods. They are overlapping and interwoven in the teaching-learning situation. However, for discussion purposes they are classified by names and labels which may not completely describe what a teacher is doing. The purpose is to provide you with a "name tag" so the discussions may prove meaningful.

Explaining and Telling

Industrial education teachers find that the method of explaining or telling is one of the most effective communicative devices at their command for presenting technical content. It is sometimes used as a long and carefully planned presentation usually referred to as the "lecture" method. On many occasions it is used for brief and spontaneous explanations in conjunction with other methods of instruction. The term lecture usually takes on the connotation of a formal presentation of subject matter in which the student is an inactive participant, a listener. The formal lecture should be used with caution, since it involves no active student participation and requires a teacher who is articulate in public speaking to keep students alert and keep the content from becoming boring and dull. Perhaps more students have been lulled to sleep by a poor lecture than any other method of teaching. The content may be exciting to the teacher, but he must put himself in the place of the student to examine how it will be received. For purposes of study, the term explaining or telling will be used here because it more suitably describes the actual use of this method.

Explanations should start with what the student knows and is familiar with and proceed toward the desired goal of the student in relation to the content. Any good explanation requires the active participation of students. This is one of the main disadvantages of the formal lecture. When you are explaining and telling during a presentation it is necessary to know if your students are with you and if you are with them. You do not know what is in their minds unless you provide for some type of interaction. They also need a form of interaction to be sure they are making correct interpretations of your explanations. This is where you immediately recognize that explaining and telling are not methods that are used in isolation. The need for interaction requires a combination with showing, doing, or demonstrating along with many other instructional media. Have you ever tried to explain something complicated to someone, like the operation of the carburetor on an automobile? Words alone are difficult to convey the message. Soon you get out a sheet of paper to sketch concepts that words alone are not conveying in pictorial visualizations to the other person. He asks questions while you respond and check his reasoning with further questions and explanations. The learning situation prompted the use of certain methods and the learning situation is underway.

Effective explaining may be used with a whole class, a small group, or with individual students. It is the way in which you organize your explanations along with other teaching that makes learning the most meaningful. Just because a student has been told does not mean that he knows or understands. Take, for example, an explanation on the principles and use of a micrometer for accurate measurement. Your explanation would probably be the same for the whole class as it would be to one student. You may have it so well organized that you feel certain every student will understand the concepts and use of the micrometer with ease. However, when you are finished a student says he does not understand why one complete turn of the thimble moves the spindle twenty-five thousandths of an inch. Another asks what the anvil is for and a third student wants to know how he can locate one-half inch on the hub. By this time you would probably be grasping for ways to help bring about these understandings. A large wall chart, an overhead transparency, a large model, the blackboard, there would have been many things you could have planned if you had realized that telling or explaining does not always work by itself. Get the point? Combining your explanations with other methods and media will help to make your teaching more effective.

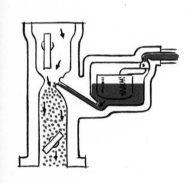

Although explanations are usually used in combination with other teaching methods, there are a number of points dealing with good explaining that are important. Perhaps the following suggestions will aid you in making your explanations more meaningful to your students.

THE BIG TWO OF SMALL WORDS. Use terminology that is "meaningful" and "understandable" when explaining. Multiporous arrangement of the lunule wires will indubitability coadunate the distributor. The previous sentence would likely have more meaning if you said that the distributor will burn out if the wires are connected incorrectly. Use words that convey the message as simply as possible and provide for the easiest understanding. Big words may make you sound intelligent but leave the students bewildered. Technical terminology is often difficult in itself and you should carefully define terms which may lead to confusion.

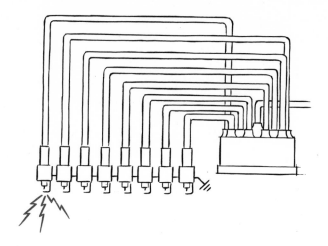

PROVIDE FOR STUDENT INTEREST. Explanations will be far more receptive and meaningful when you show the student the need for the material being explained and what outcomes may be expected.

WATCH OUT FOR WARNING SIGNS. If you start your explanations at the students' level and background of understanding, you can tie the content in with what they already know. As your explanations proceed, you will be able to recognize signs of agreement or confusion on their faces. You will know if your communication has been clear from their expressions. A good teacher will watch his class closely for such warning signs and adjust his presentation accordingly. Try to keep your explanations brief and to the point. Long explanations may lose their attention. If students are talking with each other or looking out the window, you have gone on too long and they are no longer with you.

PLAN YOUR APPROACH. Know what you are going to say and explain it in a sequential order that your students will understand. Plan to round out your topic so they will grasp the total concept when you have finished. Be brief and clear.

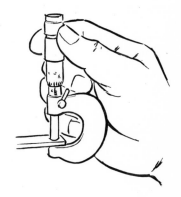

SAMPLES AND EXAMPLES. Make specific use of sample materials whenever possible. Explaining about things they can see is much more meaningful than trying to have them picture the topic in their minds. Use examples relating to the topic. These may be illustrations or instructional media of all kinds that are appropriate. An electron is difficult to explain. Yet a model of an electron as an example may well clarify your explanations. Supplement your verbal explanations with samples and examples. They are also "attention getters."

A BOX FULL OF EXCITEMENT. Many teachers keep the examples about which they are going to explain out of the sight of the students. At the correct time, each is brought into view of the class as the explanation progresses. This provides for a degree of anticipation and excitement, since they are not aware of what interesting example they may have an opportunity to learn about next.

KEEP IN CONTACT WITH YOUR CLASS. Too often a teacher loses contact with his students by doing all the talking. Plan your explanations so they will solicit questions. Let your students participate. Use questions whenever appropriate. Answer questions as they arise. Do not get sidetracked and pulled away from the objective of your explanation by unrelated questions.

MAKE APPLICATION OF YOUR INFORMATION. Follow up your explanations with student activity and performance. If you have just explained a technique, let them try it. If you have explained a process, let them go through it while they understand and are still interested. It takes advanced planning but the results are worth the effort.

EVALUATE YOUR EXPLANATIONS. Did you really get the concept across? Do your students understand what you explained in the same light that you do. Check on their performance. Ask questions to see if they understand the principles involved. See if they have gained the objective of the learning situation. This will help the students and it will also help you in making revisions in your plans and techniques for explaining.

Discussion

Discussion is a most valuable method of teaching and yet it is undoubtedly one of the most difficult to master for the ultimate in student learning. Demonstrations and explanations are easier to plan ahead and put into practice which undoubtedly explains why they are used so much. They are more teacher orientated. You do not necessarily have to depend on your students. Discussions could be classified as group "debating." It is the major form of learning in the idealistic method of teaching for it depends on the sharing of ideas, information, attitudes, and experiences. Discussion provides for everyone to "get into the act" in learning, and if it is handled correctly, results in one of the best teaching methods at your command.

It should include all of those activities which tend to develop an interchange and interaction of ideas and attitudes between teacher and students and between students themselves. The objective of the discussion should be the solution of a problem, the resolution of a conflict of understanding, or the development of a logical conclusion to a question. The attack should be the mutual interchange of experience and ideas. The result should provide for deeper insights and understandings on the part of the students. It is most important that the process of discussions includes the flow of ideas and information from student to student, from teacher to student, and from student to teacher. It is a "three ring circus" of learning. Failure on any part of the three components will result in an inadequate method. This helps to explain why it is necessary, as a teacher, to devise, plan, and study all the requirements necessary to make this method successful.

It is difficult to isolate the discussion from other methods. It should not be limited in any way by denying the use of any technique that may prove advantageous. On the other hand, the discussion must proceed in an orderly manner, similar to a well supervised debate, in which the conversation among students and between students and teacher is always directed toward obtaining a definite objective. Heated arguments only hinder this learning situation and have no place in its employment. It should be well understood that the discussion is a group thinking and sharing process in which the teacher plays the role of both a participant and leader. Now let's consider the appropriate uses of the discussion method in industrial education classes.

When to Use Group Discussions

Group discussions are certainly not adequate for all learning situations but they are advantageous in many instances. For example, if you had the objective of teaching your students how to correctly grind a twist drill, a group discussion would hardly prove a sound method. If your plans called for students to solve a problem dealing with the heat treatment of metals for particular applications, the group discussion may be the most successful. The mental or manipulative skills involved with the learning activity and the principles to be learned in order to obtain a desired goal will usually serve as a guide in determining the effective use of discussions. In general, the discussion is of greater value in dealing with problems that may have many possible solutions or are of a theoretical nature. A few ideas concerning the appropriate use of discussions follow:

When it is necessary to apply technical theory or procedure to practical applications.

In specific instances, to clarify and sum up information on broad topics.

When problem solving situations arise and the possibility of many solutions exists.

During times when planning of student activities, assignments, readings, and experimentation involve many proposals.

To be used for evaluative purposes concerning student activities and laboratory management.

As an aid in sharing information in the form of a seminar in which students discuss their own technical activities.

To prepare students for fundamental activities and instruction which will follow throughout a course.

Planning Group Discussions

It is most effective to have any group discussions planned in advance in order to keep the class directed toward the objectives to be achieved. In planning your group discussions, there are three main points to consider and follow: First, be sure you are starting the discussion with what the students already know so they will have a background for active participation. Second, you should plan where and how the discussion will take place. The class may be seated around a planning area, in front of a teaching center for the use of instructional media, or standing around a machine or other equipment. Third, the discussion topic should be of importance to the total group if it is to be of value to each class member.

You should remember that a discussion will only be as good as the leadership provided by the teacher. Any breakdown in a group discussion fails to bring about the desired behavior changes of the students and is apt to waste considerable time. As a group discussion leader, you have a number of responsibilities to keep the discussion flowing smoothly and in the direction intended. The discussion should start with some common experiences the students possess. These experiences may have been acquired the previous day or at any other time in their technical learning environment. They are your starting point for beginning discussion. Here is where you need to take command and yet appear to remain in the background. Do not let a planned discussion turn into a lecture. Start it off with a "bang" to whet student desire to participate. Your best resource here is your ability to throw out questions which will cause your students to further question one another and explain their points of view. Once the group has established the tie-in with their previous experiences, the discussion may flow toward the desired objective without your participation. On the other hand, no two groups are the same and you must be ready to give the discussion direction if they get off the track, direct questions that will keep the discussion going if it bogs down. If it appears the discussion is wasting time, take a more active part by giving an explanation or demonstration.

Teaching industrial education classes by the use of group discussions from time to time also serves as an excellent means of evaluation. You may be able to determine the effectiveness of your instruction, student progress, student development of thinking patterns, and attitudes during a discussion. The student, likewise, may be able to evaluate his own progress in light of the thinking of others in the group compared to his own. Give each student an opportunity to test himself by questioning others in the group and allow them to evaluate each others performances.

Advantages for Industrial Education

It is unfortunate that the group discussion method has not been used more in industrial education classes. Perhaps the traditional concept that technical courses are cut and dried, consisting of only skills and knowledge to be passed on to the student, has limited its use. However, technical courses in the modern industrial setting include such vast scientific and theoretical information along with mental and manipulative development that the discussion method becomes all the more important in bringing about the understanding of materials, equipment operation, servicing, and production. Here are a few suggestions for taking advantage of planned discussion in your technical classes:

1. It helps to develop a cooperative attitude among members of the class and increases student interest.

2. Discussions stimulate student thinking about technical problems because they know they are expected to participate. You should bring up problems that they recognize are important and clearly define the scope and limitations expected in the discussion.

3. You have the opportunity to correct any misconceptions that students may reveal as you try to get all the students to participate. Use the experiences and skills of the total group to help overcome misunderstandings.

4. Plan discussions around complicated machinery, equipment, and processes which will provide for an interchange of ideas, questions, explanations, reactions, and understandings. You then have the opportunity to summarize important points and conclusions reached by your group.

Demonstrating or Showing

Industrial education courses are loaded with technical equipment, tools, new materials, machines, and processes which require the method of demonstrating or showing to assist the student in better understanding and using such equipment. Nowhere else in our society or schools is there more need or room for this method of teaching than in courses dealing with industry. Practically everything in a technical laboratory, from a hand drill to an electronic computer, challenges a teacher to provide effective demonstrations to bring about student understanding.

"Show and tell" sound like the ingredients of a Monday morning elementary school group when weekend experiences are proudly presented to the class. Yet for hundreds of years, even before the beginning of the apprenticeship system, the process of showing someone how something works and describing its operation has been an essential feature of education. We also fail to realize sometimes that over these hundreds of years most all of the showing and telling dealt with some phase of industry. Its importance is even greater today than ever before since many of the demonstrations you will be required to present will be more complicated than the spinning wheel or the hammer and horseshoe.

The purpose of a demonstration is to illustrate how a process, procedure, or experiment, is to be done so you may aid the student in acquiring the knowledge or learning the skill. It is one of the most effective teaching methods used in industrial education courses. The demonstration is given to show the student exactly what is to be done, why it is done in a certain way, how to do it, and how to apply the skill or procedure that has been presented. A demonstration may be used in many ways and for many purposes. It may be used to illustrate and teach a concept or principle and show the application of the principle. It may be used to establish a problem solving situation and post a challenge for student learning. The demonstration may also serve to show a simple manipulative skill which would be followed by student performance. Your demonstration may also serve as a standard for evaluating the performance of students as they pursue an activity. In any of these applications, the demonstration should always be used in conjunction with other teaching methods, primarily with an explanation. Showing alone is not enough. An explanation of what is being performed is always necessary. Further, including other teaching methods and instructional media often makes the demonstration more effective as a teaching device.

Planning Your Demonstrations

Just as important as the demonstration itself is the preparation and planning that goes into it. A sloppy demonstration will probably produce sloppy performance. The demonstration should be given with enthusiasm, skill, and determination. It is a show and you are the main actor. Your performance will have a considerable influence on the attitude of your students. You are actually in competition with other communications media, such as films and television, when you present a good demonstration. This is not to say you need to take acting lessons, but it does indicate some of the qualities a teacher should possess for giving effective demonstrations. A good speaking voice, tactfulness, a neat appearance, and a bit of humor go a long way in putting your demonstration across. Concentrate on these as you plan and give demonstrations to encourage student interest and attention. These are prime requisites if the demonstration is to be extremely meaningful to your students.

Plans for the technical demonstration should be carefully made in advance. Lack of preparation will be all too apparent if things do not work right or you become confused during the presentation. Here are some helpful suggestions for planning and making your demonstrations:

1. Decide exactly what you want to demonstrate so you will not provide too little or too much. Try to put yourself in the place of your students to see what they need to get from your demonstration to gain the necessary understanding or develop a skillful performance.

2. Practice the performance you will present to be sure that it is technically correct. The skill with which you handle equipment and perform operations will set a standard of performance for your students. Practice what you want your students to do by doing it yourself until you have mastered all techniques. Your demonstration should be performed with a greater degree of skill than you would expect from your students. They may pass you up some day, just as a professional quarterback is able to out-perform his coach, but it did not start out that way.

3. Plan how you will present the demonstration and be sure that everything works so you can carry it through successfully. If you have tools and equipment to be used, be sure they are ready and in good operating condition. The show is over if you have to say, "Wait a minute while I go sharpen this cutter bit." Try to anticipate difficulties or questions your students may have. Define your objective and plans for what you expect the students to learn from your demonstration and make arrangements for students to participate where it is appropriate. They love to "get in on the action."

4. Give your students a chance to anticipate what the demonstration will be all about. Let them know in advance and explain the importance of the material to be learned. A "sneak preview" may be interesting but will it really sink in.

5. Make arrangements so that all your students can see and hear. Since there are so many forms of demonstrations, make your plans in advance so your students will not be continuously moving around to be able to see what is going on. If you are demonstrating something that is portable, say the operation of small gasoline engine, use a table around which students may be comfortably seated. When parts are too small to be easily seen, make use of models and other teaching media which will emphasize important points. A demonstration center provides ideal learning atmosphere since all of your teaching media are readily available. If you are making a demonstration on large equipment, try to arrange for all students to see and hear and take the necessary teaching aids with you. Stand so you do not hide what you are showing. Your students learn by seeing what you are doing.

6. Give your demonstration and make it exciting and enjoyable. You have planned and practiced, prepared materials, let your students know what to expect, now demonstrate how it works. Use all of your teaching skills. Ask questions as they appear appropriate, summarize points of special importance, go slow enough so all students will understand complex operations, and make your explanations clear as you proceed. Demonstrate and explain at the same time. Give illustrations of the "why" of what you are doing as you go through each step. Review any steps in your demonstration carefully when questions arise or doubt is expressed on a student's face. Return questions to see that a point is understood. Remember, a good demonstration is one of your best means of communication for student understanding and performance.

7. The demonstration is perhaps the best opportunity you have to put real meaning into safe operation and use of equipment. Practically every procedure, process, or operation in industrial education courses has some potential for accidents or injury. Emphasize any special precautions. Show how to handle equipment safely and point out the possible dangers of misuse or carelessness. Set an example for your students. You are the one person they will most likely imitate in performance of manipulative operations.

8. Have your students really learned from your demonstration? A follow-up and evaluation is an ongoing process in which you should carefully observe their procedures or performance. When students have had an opportunity to perform, you may have to repeat some aspects of your demonstration individually or to small groups. If you have shown a student how to do a pictorial sketch or a proper cutting process, only his correct performance will prove that your demonstration was successful.

The effective demonstration has many advantages as a teaching method. It saves valuable learning time in the sense that understandings may be obtained much quicker than through just verbal explanations. It is appealing and arouses interest when it is well planned and performed. The advantage of appealing to most all of the human senses is overwhelming. Besides hearing and seeing, the student often has the opportunity to feel and touch, and in some cases learn the odors and tastes of various materials. The demonstration gives meaning to words and understanding to concepts. Plan, organize, and make demonstrations educationally meaningful. Take advantage of this most valuable teaching-learning method and give your students a real opportunity to learn.

Performance

Performance is what learning is really all about. The end result in all teaching-learning situations is based upon the student's ability to do something well - - to perform. Our educational system is based upon the development of mental and manipulative powers in such a combination that students may be able to better perform in specific and broad situations. Just thinking without action has little meaning. The same exists for doing or manipulation without mental direction. The concept of "learning by doing" should take on a significant meaning for you as a technical teacher as you work with students in their performance of activities.

Earlier educational theory did not place emphasis on performance as it is conceived today. So-called cultural education placed the emphasis on the development of the students mental powers. You may recall the inquisitive child who asked his friend, "What does your father do?" The friend replied "He's a philosopher." Yes, I remember that," continued the child, "but what does he do?" "Oh, he just sits around and thinks," said his friend. Interesting, but you well know that just deep processes of thought does not make a philosopher. His thinking must result in action or performance, or it is useless.

Can you imagine a person trying to learn how to play baseball without a ball or swinging a bat? It would be just as absurd to try to learn how to water ski without a boat and tow line, or learn how to play chess without a board. It's performance that counts, along with mental development. To learn how to drive a car without getting behind the wheel is almost impossible. To learn how to drive a car by physical trial and error is dangerous. Driver education in our schools makes a vivid illustration. Students have an opportunity to do all the types of things they have been studying. Explanations are given about the car and its operation, reading and activity assignments are made, demonstrations are given on preventive maintenance and operation, audio-visual materials clarify traffic laws and safety, and finally - - the student performance.

Industrial education courses thrive on student performance as an outstanding feature of laboratory activity. Experimenting, constructing, processing, manufacturing, testing, are but a few of the formal names given to student performance. These types of performances engage the student in the most active part of learning. However, they are not just a "do it yourself" activity. They involve as much or more preparation as a demonstration or explanation. In fact, they are a part of and the result of your many teaching methods. Performance requires active participation on the part of the teacher as well, in guiding and supervising the student activity.

During the performance phase of a learning situation, the teacher uses any of his resources in methods to help bring about any concept, skill, or understanding. While watching student performance it may be necessary for you to make further explanations or "on the spot" demonstrations. You may have to ask questions and answer questions to guide his performance. For example, let's assume one of your students is preparing a metal sample for microscopic examination. You have gone through a complete presentation on this activity, but you find he is still having difficulty. He may not ask for help. This is where your observation of performance becomes so important. You become quite sensitive to inadequate performance, so you ask what difficulty he is having. He may not be aware of having any. Soon your discussion leads his thinking back to something he missed or did not understand. So you show him again. He adjusts the microscope and you ask what he sees. He explains and you assist him in locating what he should be seeing, and on and on it goes. The interaction of student and teacher during performance is the key to good activity instruction.

The student activity or "doing" may occur in many forms in a technical course. It should involve carefully planned activities which bring out both physical and motor responses as well as reasoning, problem solving, writing, or creative thinking. One of the poorest forms of performance activities is that in which the student blindly follows a set of established steps of procedure without really understanding what he is doing. You should be sure that any activity in which you have students engaged is planned to guide them but does not give all the answers. It is very possible for a student to complete a job quite well and yet be unaware of the concepts he was supposed to have learned. Work carefully with your students during performance activities to be sure they are learning. "Learning by doing" involves just what it says, "learning" and "doing." When a performance only involves the "doing" phase, the prospects for learning become very dim. As a teacher, you must use every technique available to make sure the student understands. Then student performance can be highly effective.

Using Your Performance Activities

You have had an opportunity to study and plan teaching activities within your technical area of interest. These activities have been discussed on the basis of providing the student with learning opportunities covering a broad spectrum, from teacher dominated activities to completely student centered activities. The effectiveness of any of these activities will become apparent as the students engage in performance in a laboratory situation. Your responsibility now is two-fold, to work closely with your students as they carry out their performance activities and to evaluate their performance in light of the short and long range goals. The following suggestions are presented to assist you in this phase of your technical teaching responsibility:

1. Prepare your students for their planned performance activities as carefully as possible. Make your explanations, demonstrations, or whatever necessary presentations, clear and understandable. Let them know what is expected in their performance and why. Careful planning with your students helps to eliminate those constant questions of, "What am I supposed to do next," or "I don't understand how to do this operation." Questions like these can never be completely eliminated, but you can keep them to a minimum by taking the time to clearly outline the activities. Your planning time with your students should also provide for motivation as has been discussed. Get them ready, interested, and excited by helping them see the need for the activity and what the objective is.

2. Have your supplies and equipment ready. Make plans so your students do not have to keep asking you for materials or supplies. This is valuable laboratory time for you and your students. Do not waste it getting out materials. This should be done before your classes begin.

3. During laboratory activity, encourage your students to perform to the best of their ability. They need your assistance to work with precision and accuracy. Give constructive criticism to those who need help. Ask questions and try to identify problems as they arise. Encourage the slower student with patience and understanding and confront the better students with more challenging problems.

4. Evaluate student performance all the way along the line. Help all your students in making self analysis by asking questions. The better he is able to see and understand his own mistakes, the more aware he will be of improving his performance and be directed toward his goals. Your observation of what your students are doing and how they are getting along with their work is the best measure you have in assisting them toward the objectives of the course.

Selecting Appropriate Methods

Each industrial education course offers so much technical content and information that it is not always an easy task to select appropriate methods for presenting material. Some guide lines do exist, however, that help in selecting teaching methods for certain types of content. There is no bag full of "tricks" that will suffice for all situations. It is a matter of identifying the understanding, skill, or concept to be learned and selecting a method which will most likely bring about the desired results. But how do you know what method is appropriate? You must search, analyze, and try out various methods until you get the "knack" of associating methodology with content for optimum learning. A few things are obvious. Any one method of teaching, say for example the demonstration method, will not be satisfactory for all learning conditions. For example, you plan to teach the concept of weights and measurement using a balance. After thinking it over, you decide to attempt to bring the concept about by an explanation. The chances are you will not get the results you had planned when you later observe students using the balance. It is difficult for students to obtain an understanding of how to set up and manipulate the balance by just hearing the process described. Can they actually picture the whole process in their minds? Undoubtedly the use of a number of methods would be more appropriate. A demonstration of the operation of the balance, using a few sample materials, along with an explanation of the principles of calculating weights would probably put the point across much quicker and with better understanding. Add to this some planned questions, a sample calculation on the chalk board, and student discussion and you may well have an interesting and understandable presentation. Many methods at work usually reduce learning time and increase understanding.

Let Your Content be Your Guide

The outline of your course content and your units of instruction should provide the necessary guide lines for planning the methods you will use. Your planned learning activities should also be contemplated, for they will require certain instruction necessary for student performance. The nature of your technical content will call for specific methods in many instances. A course in electricity would certainly be taught with emphasis on some methods which may not be used in a drawing course. The content in electricity is more theoretical in nature and would require different methods than those used in drawing. Different in the sense that the combination of methods and instructional devices relate to the concepts being taught. Certainly all methods would be used in both courses. It is "how" they are used that is of utmost importance. It would be hard to imagine how a demonstration or lecture could bring out the concepts of electron flow through a circuit without other teaching media

such as diagrams, overhead transparencies, or animated films. On the other hand, in the course in drawing, mock-ups, models, plastic boxes, and other devices may better be used with explanations for various topics.

These, of course, are general considerations which should apply to some degree in any technical course. Your major problem is to analyze each topic of your teaching unit and plan the best way in which you can "get it across" to your students. It should be tied in with previous topics and lead to the development and understanding of topics to come. As you analyze each topic, there are a number of questions you should ask yourself as to the approach or methods you will use in making that topic the most "teachable" and the most "learnable." Every topic is somewhat unique in itself and certain aspects of it will help you plan your approach. The following questions should prove helpful in planning your methodology for any topic of content:

1. What communication method will best help the students understand the topic?

2. What instructional media would aid in the presentation?

3. Is there anything about the topic which would require something to be read from resource material or a hand-out?

4. What will be the most difficult aspects of the topic for the students to understand? How best can these be presented and explained?

5. Is there anything about the topic which could best be learned outside of the laboratory, such as a field trip?

6. How best can the students become "actively" involved with the topic?

7. Are the necessary materials and equipment available for a presentation, or must some things be constructed?

8. Will the presentation lead into an understanding of planned activities for performance?

If you ask yourself these questions, they will certainly assist you in planning the finest combination of methods and instructional media necessary to do a good job of teaching. Try all kinds of combinations and before long you will find it much easier to plan what methods to use and when. Just one caution; do not get "stuck" with one approach which you begin to apply to everything you teach. Always be thinking out new methods and continually be receptive to making changes when it appears it will make your teaching more effective. Students, the industrial education teacher with his vast resources in methodology, and the wide world of industry are an unbeatable combination for the ultimate in learning.

Evaluating Your Teaching Methods

How good a teacher are you? Are you using all your resources in teaching methods? Are you selecting the most suitable method for the particular type of content you are presenting? Do your students seem to be "with you" when you are making presentations? These and similar questions can be answered best by the attitudes, abilities, skills, and understandings your students have gained toward the objectives of your course. Your own evaluation of your methods of teaching is directly related to performance and application of principles learned by your students. However, student evaluation is not the whole story in taking stock of your teaching methods.

After every presentation or demonstration you should quickly review just what you did, what you used, and how everything went. Perhaps student questions arose that would not have been necessary if you had used a different technique or another type of instructional aid. Your observation of student performance after presentations will also aid in evaluating your methods. Why were some students having so much difficulty? Ask your students. Many times a student will indicate things which will be very helpful in making changes in your approach. Such things as, "If I had only seen a complete diagram of what the inside of the machine looked like, I would have better understood the process," or "If you had shown us the slides before and after we did our heat treating, I think it would have been easier to understand." Take student reaction to heart and adjust your presentations accordingly.

Evaluation and reorganization of your methods of presentation is an ongoing process. It appears you could never run out of new ways of continually improving your instructional techniques. Take account of what you have done, student reaction, new materials you could prepare, and make the necessary reorganization of your methods of presentation. You are then on the road to becoming a better and better industrial education teacher.

Review Questions - Chapter 8

1. Why is problem solving considered to be the framework for methodology?
2. In what ways may students be helpful in assisting you in evaluating your teaching methods?
3. What are the individual steps in the overall problem solving learning situation?
4. Why should a student and teacher work together on a problem solving situation? Why shouldn't the student just be left on his own to attack the problem?
5. What are your resources in teaching methods? Are there really any limits to your resources? Why?
6. Why is it so difficult to draw a sharp line of demarcation between different teaching methods?
7. Why are industrial education courses usually unique compared to other courses in relation to instructional methodology?
8. What are the advantages of involving your students actively when you are giving explanations?
9. What makes the method of discussion one of the most valuable teaching techniques at your command?
10. What suggestions do you have for the appropriate use of discussions in technical courses?
11. What responsibilities do you, as an industrial education teacher, have as a group discussion leader?
12. Explain how the discussion method may serve you as an evaluative device of your students.
13. Why is the good demonstration even more important today in technical courses than it ever has been?
14. What are the most important features in the planning of a good technical demonstration?

15. Give a number of reasons why industrial education courses thrive on student performance.
16. In what ways does observation of student performance assist you in helping students learn better?
17. What are your two-fold responsibilities in relation to student performance activities?
18. What are some of the guiding factors you should follow in selecting appropriate methods for specific topics to be learned?
19. Explain why one specific teaching method can seldomly be used effectively in any learning situation.
20. Why does the method of appealing to as many student sense perceptions as possible usually result in better learning?
21. Explain how your course content can help serve as a guide for the selection of teaching methods.
22. Why does it seem so easy for some teachers to get stuck on certain methods and neglect others? What can be done to remedy such a situation?

Suggested Activities

1. Select a topic from your content in your technical area of interest and make an outline of how you would present the material in a discussion situation. Remember that every topic may not be advantageously presented through a discussion so select your topic carefully.

2. Make a listing of your resources in methods in teaching a unit from your course outline. Plan your listing so that each major topic makes use of the best methods you could develop and present. Use reference material and the Chapter on Instructional Media for assistance.

3. Select a technical magazine or industrial publication that contains some modern process, new materials, or other pertinent information not in your course outline but of importance to your technical area. Prepare a presentation, using any combination of methods, on a specific article or information from the publication which would be valuable to your course.

4. Make up a demonstration outline for a selected topic in your technical area. Pick a modern process or piece of equipment for your demonstration and include the necessary teaching media you feel would most likely help put the demonstration "across."

5. Using one of your laboratory activities, make an outline of the explanations necessary to get your students ready for performance of the activity.

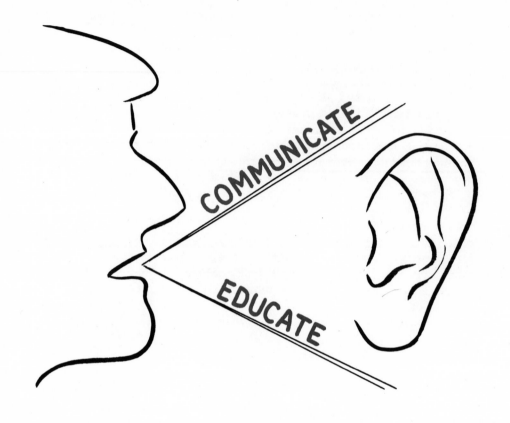

Chapter 9
EVALUATION IN
INDUSTRIAL EDUCATION

A broad concept of evaluation in industrial education courses is one of the most significant attitudes that a teacher may acquire. It is not easy. It is also not an attitude unrelated from all the other phases of teaching.

Appropriate and meaningful evaluation must be thought of as an integral part of the teaching-learning process. It must also become an integral part of the teacher's attitude and personality. The concept is somewhat like the attitude of an excellent driver toward safety on the highway. He is so aware of traffic safety and potential hazards that he develops a "built in" mechanism for evaluating each and every situation. He appears to see the whole process from the inside out and is able to react quickly. The evaluative perception he possesses is an ongoing process, not an "after the accident" estimate of what really happened. So too with the good industrial education teacher. He seems to have "peripheral vision" as to how students are performing, how they feel about things, what they are thinking about, and how their mental development is shaping up at all times during a course. He is concerned about their attitudes as well as their understandings. He is concerned about their ability to perform as well as their performance. This is the nature of the evaluation process toward which you should direct your attention. A broad concept which prevails during all learning and the interrelated principles of individual student achievement. As you begin a study of student evaluation, think in terms of what a problem in magnetism does to your students rather than what your students do to a problem in magnetism.

How good is your teaching? How well are your students learning? These two considerations are actually the foundations on which evaluations are based. They are stated separately and yet they are really one. For in all aspects of industrial education programs, the purpose is to promote student learning toward desired goals. Your teaching and student learning are the active methods through which these goals are obtained. An evaluation of what the student has learned, the performance skills he has acquired, and his mental development in identifying and solving problems reflects the effectiveness of your teaching and the ability of students to learn. It's that simple; on paper. However, a broad concept of evaluation is quite complex. Within this complex concept lies the two basic considerations of "what" and "how" to evaluate. First of all, let's take a look at the place of evaluation as part of the teaching pattern.

Evaluation As a Part of Teaching

Some aspects of evaluation should be integrated into all phases of your teaching-learning pattern. Your evaluation processes actually begin when you provide clearly defined objectives for your course. You are giving your students direction as to where they are going and what desired behavior changes are expected. Your evaluation of behavior change is based on what you and your students expect that behavior change to be.

Your evaluation techniques should be used to assist students to better understand your presentations and explanations. Make your evaluative tools serve both you and your students. When you are giving explanations and demonstrations, you should be constantly checking your students progress and understanding by asking questions, answering questions, and observing their performance. Any teaching methods you use, or any student activities being performed, should include ways in which you can better understand how your students are learning... how their behavior is changing in light of the goals of your course. Do not just give a demonstration or make a presentation assuming that when you are finished your students will all understand and be able to perform. That would be guessing, not evaluating. Watch your students when you are presenting material. Observe their expressions, be alert for any indications of puzzlement or misunderstanding. Correct any situations immediately when you see you are not getting the point across. This is in-process evaluation, checking your teaching methods and student understanding as the learning situation progresses. Your students can give you a lot of answers concerning how well they are learning as well as what they are learning.

O
B
J
E
C
T
I
V
E
S

Try to tie your evaluation in with student motivation. The student likes to know when he has done a good job. A pat on the back or an encouraging word provides further incentive. He then knows that you feel he has done a good job and he knows that he has been evaluated. Not formally with a check mark or letter grade, but his experience of success provides a satisfaction toward better performances in the future. He must also be aware of his progress toward his overall goals. He must know whether his efforts are in the right direction, if he is succeeding or failing. Your teaching evaluation helps him when he fails by letting him know why he has failed and how he can improve his performance and understanding.

Laboratory evaluation is also an essential part of your teaching. Although you are not formally presenting demonstrations or instructions, you have perhaps your best opportunity to teach and evaluate simultaneously. Observation is the most valuable tool you have at your command. As you observe student performance, point out what the student is doing correctly and what he is doing incorrectly. If he is having difficulty, show him how his performance can be improved. Make him aware of self evaluation so he may better understand his errors and make his own corrections. The threads of teaching and evaluation should be woven through technical performance. The student needs to know where he is going, how he is getting along, and how well he did. His ability to learn and perform will depend a great deal on his knowledge of his progress. You must assist in evaluating his progress. You must also assist in helping him understand and evaluate his own progress. The good industrial education teacher cannot separate teaching from evaluation, nor learning from instruction. Now the cycle is complete. You have gone from teaching to learning to evaluation and have found that they are all interrelated and take place at the same time.

Ongoing Evaluation Process

The broad concept of evaluation has illustrated that it is not just a matter of giving tests to find out what a student knows. Nor is it just a matter of grading projects to find out what a student can do. These will not get the entire job done. If you can develop the realization that educational outcomes vary for each person, and that it is the student's behavior rather than the technical content to be evaluated, you're off and running. This statement must be qualified. Not whether the student is being polite, chewing gum, or talking too much. Changes in student behavior include "all" of his mental understandings and manipulative developments in direct proportion to the objectives of the course. You are evaluating the student's growth and development as an ongoing process of learning.

When you are able to achieve this attitude toward evaluation, you will find that the education of your students rather than the teaching of beloved content becomes your major concern. The content won't lose any of its fascination and delight. It will be just as important. Probably more so, because of the rapid technological changes. WHAT THE STUDENT DOES WITH THE CONTENT SHOULD BECOME YOUR MAJOR CONCERN. In this way, only an ongoing process of evaluation will make it possible for you to assist students toward their goals of learning.

You need devices and techniques for evaluation to make this ongoing process possible. You have some of them, and you will have to work hard to revise and develop others. Progress charts, cumulative records, conscious observation, student self evaluation, cooperative teacher-student evaluation, student-to-student evaluation, and all of your performance tests and examinations should be exploited for all their potential. You will then be getting closer to the real meaning of evaluation; the continuous viewing of the learner's behavior change by both student and teacher.

Let the Student Know What to Expect

Any industrial education course should be directed to the planned objectives and selected content of the teacher. In turn, it is the teacher's responsibility to provide his students with all information concerning what he expects them to learn, to be able to perform, how to conduct themselves in class, and the manner in which the course activities will be carried out. You must let the student know what to expect in terms of teaching, learning, and evaluation. He should be made well aware of how his learning and performance will be evaluated. If not, you are not providing him with the guidance necessary to let him know exactly where he is going. Any lack of communication between you and your students as to what you expect of them will certainly make evaluation difficult and ineffective. Such comments from students as, "I didn't know you were going to grade us on how fast we worked," or "You didn't tell us we could get a higher grade if we made a drawing instead of a sketch," should never come up unless a student has not paid attention. Your evaluation will be just as meaningful as your student's understanding of what is expected of them.

Likewise, you should let your students know exactly what they can expect from you. There is no room for "trick plays" in your teaching. Trying to catch one of your students off first base is no way to motivate him to hit a home run the next time he is at bat. Secure their confidence. Tell them exactly how you will play the game and decide on the rules together. Industrial education is one game in which both teams can score high and yet both win. Some factors which should help you in letting the student know what to expect are as follows:

1. Discuss the objectives of your course with your students in a clear and simple manner.

2. Explain precisely what you will be observing during their study, activities, and performances.

3. Tell them how, when, and where you will make specific evaluations.

4. Explain homework assignments clearly. You will not be there for them to ask questions, so be sure they understand exactly what is to be done.

5. Incorporate evaluation procedures into your presentations and student activities. Discuss these with your students at the time they are being used.

6. Let your class know that their questions are important and will assist you in teaching and evaluation.

Characteristics of Evaluative Devices

Evaluation is somewhat similar to teaching methodology. There have been so many names given to various evaluative devices that it often appears they are used independently and only for specific purposes. This is seldom true. Yet, like teaching methods, they are defined separately for purposes of discussion. During actual use they are usually interwoven and overlapping to provide the best scheme for total student evaluation.

A program of evaluation includes all the methods and techniques that an industrial education teacher uses to keep his students and himself informed as to the progress they are making. These evaluation devices are not necessarily printed to require a written answer. In a broad concept of evaluation, they would include any device that appeals to any combination of sense perceptions of the student. They include progress charts, oral questions and answers, conscious observation, examinations and review of examination results, projects, student activities, teacher rating devices, personal interviews, homework assignments, text assignments, and class participation.

The characteristics of any good evaluative devices are that they have the potential for helping students learn and provide the kinds of information you desire. They should be kept as simple as possible and yet yield the necessary results. Evaluation of understanding and performance in terms of student behavior change is the real goal of the industrial education teacher. You should be able to choose the evaluation device best suited for the particular purpose to be served. You must also have command of a wide range of evaluating methods and techniques to make your evaluation as meaningful as possible. The characteristics of evaluating devices which prove the most beneficial can best be described under the following headings of comprehensiveness, reliability, and validity.

Comprehensiveness

Any evaluative devices should be comprehensive enough to adequately cover the subject matter or concepts to be learned. They should be directed toward the objectives you are attempting to evaluate and cover enough factors to provide the information you want to receive. It is often difficult for you to hit on a happy medium. One or two questions over a lot of material may not give you an adequate sample. On the other hand, if you attempt to cover every minute detail with some evaluative device, you may find it too time consuming, awkward, and complex to easily handle.

The comprehensive characteristic of serving the desired purpose and still be quite easy to administer and score is also important. How much time can you afford to use from instruction and student activity for evaluation? You must make this decision according to the methods you use and the value of the evaluation scheme as it relates to student learning. Take the opportunity to check periodically just how much sampling is necessary to obtain the desired results. Evaluation should be comprehensive enough to obtain desired information and yet restricted to keep students from becoming bored or agitated. Over-evaluation interferes with student interest, motivation, and desire to learn.

Reliability and Validity

Two terms dealing with evaluation that are difficult to "put a handle on" are RELIABILITY and VALIDITY. This, of course, denotes that you are thinking in terms of a broad concept of evaluation as discussed. When complete emphasis is placed on reliability and validity, the teacher often misplaces the real meaning of evaluation. "Who wrote the graphic arts textbook by J. L. Smith?", is the first test question. Reliable? Absolutely. Valid? Undoubtedly, if your goal is the correct answer. Meaningful? No need to even discuss that since you get the point. It sounds absurd, but so are many other poorly planned evaluative devices. Now take a look at the deeper meanings of reliability and validity.

The concept of reliability refers to the consistency of our evaluative device. Can you depend on your methods and devices to produce somewhat consistent results? This will depend on how well you give directions, provide adequate sampling, check and recheck previous responses, design your test items, make allowance for objective scoring, or direct your evaluation toward your objectives. No evaluative device is 100 percent reliable. There are too many human factors involved for every student to interpret all test or evaluative devices the same way. Your care in the preparation of rating scales, test items, or any other device will prove to be best measure of reliability. The big question is how reliable you are in taking the time necessary to design respectable evaluative devices.

The concept of validity is similar. Any evaluative device is as valid as the results it indicates compared to the results it was planned to indicate. No more or no less. Any evaluation device, therefore, has a degree of validity. Some parts of it may be more valid than others, since they have told you what you planned them to tell you. When you plan a device or test item to reflect a particular concept and it gives you something entirely different in return, you are receiving an unintended false picture. It requires study, practice, and an attitude of receptive criticism from your students to obtain a respectable degree of validity for your evaluations.

Another aspect of validity is its relation to the type of evaluation device being planned. The more factual or objective the device, the more you should watch for indications that it will do what you want it to do. See if the results will really indicate what you are trying to find out, or are you receiving something else. Be sure you evaluate the device to see if it really includes the vital things necessary for a student to accomplish a task.

Say you ask a student, "How are you coming along with your lettering?" He may reply, "I think it is improving a lot. See how much better these last two drawings are. I've been practicing a lot." The degree of validity is not so meaningful. You could ask the question in many ways and receive the same response. Yet you are most certainly making an evaluation. Reliability and validity are very important characteristics with which you should be concerned when planning evaluative devices.

Subjective and Objective

There are a number of ways you will hear subjective and objective evaluations defined. These two terms mean many things to many people. You will note that some people refer to objective tests as those which require a memorized answer and subjective tests as those where the student may express his own viewpoint. Others indicate that all tests are objective to the degree that the personal judgment of the person scoring the test does not affect the score. In other words, if a number of teachers grade the same test and get the same score, the test would be considered highly objective. If these same teachers arrived at different scores, which would often be the case in essay type tests, then it would be considered subjective.

Student interpretation of the evaluation devices or items is another aspect concerning objectivity and subjectivity. For example, a poorly planned evaluative device may solicit many different responses from many different students. Even though the students know the information, are able to make correct decisions or applications, and would take the action desired by the teacher, they may come up with many different responses. This is usually due to misleading instructions, ambiguous questions, poorly worded statements, or incomplete information. It is plain to see that such a situation becomes quite subjective, for it is the student's interpretation of the items that you would be evaluating rather than his understandings and knowledge. The results are meaningless. There is no basis for comparison of what the student knows with his response to a poorly stated item.

Although arguments will probably go on eternally as to the merits of objective and subjective evaluation, a broad concept of total evaluation would include an across-the-board coverage of both. In order to fulfill your responsibilities in helping your students learn, you should give them an opportunity to respond with a factual answer, all the way to a personal viewpoint or opinion. The main point is that your evaluative devices should be consistent with what you expect for a response. If you want a student response filled in a blank to be "copper," do not phrase your question to mislead him in such a way that he will write two paragraphs on the smelting of metals. You are being objective and your student is being subjective. For evaluative purposes, the two do not mix.

As you study the preparation and application of a wide range of evaluative devices to be used in industrial education course, keep in mind the value of meeting the needs of your students. The concept of subjective and objective evaluation should fall into place and take on a new perspective.

Evaluative Devices and Test Selection

The preparation of evaluative devices and selection of adequate tests to meet the needs of your students is vital. They provide the important answer to the question, "How are we doing?" If you are to rate your students reliably on behavior changes, performance, and achievement, all of your teaching methods and evaluative devices must differentiate. They must show even small but significant differences in both mental and manipulative achievement. Your evaluation must cover all levels of difficulty in order to obtain a desired spread in results. In the final analysis, you will actually have a profile of each student's development while under the supervision of your instruction. No two students will end up the same on such a profile, although it may appear as such when it is necessary to assign grades. In order to make your selection of evaluation devices the most meaningful, you must remember that the prime function of evaluation is to locate the ways in which the student's performance and understanding can be improved.

Evaluation for grading purposes alone is like kicking the tires on used cars to find out which will be the best buy. The stinging question is, how do you evaluate a group of students, each of which is so different from one another in mental and manipulative ability, physical coordination, interest, and desire? Do you set standards for the class? Do you evaluate each student individually on his own growth and development? Should a handicapped student be evaluated in the same manner as others? This is where your broad concept of evaluation can give you some definite direction. You will have a bounty of information about each of your students to help make these decisions when you approach evaluation with an arm load of different types of devices. One final point to keep in mind is that you must get to know your students. You must know each student as an individual, what "makes him tick," what problems he may have, and what personality factors may affect his learning. If you do not get to really know your students, your evaluative scheme may have many holes in it.

Let's assume on every one of your "best" tests a certain student does very poor. You have made your judgment. Oh, but you did not know him well enough to realize he had reading problems. Would that make a difference?

Turn your attention now to a discussion of a wide range of evaluative devices in order to gain a better understanding of what to select for specific purposes.

Evaluation by Observation

You sometimes hear teachers say that if they have enough time to watch students in action, to observe their performances, they can make a very accurate evaluation of their attitudes, understandings, and abilities. They may not be too far off the track. Much depends upon how keenly they have whetted their

senses to a systematic analysis of student behavior. It is felt by many that the development of "conscious" observation on the part of the industrial education teacher is a real break-through in modern concepts of evaluation. Thus, conscious observation becomes a real interplay between teacher and student. During the course of study, this evaluative ability is slowly taken on by the student as well. He then begins to assume some responsibility for self evaluation and takes an active part in the evaluation process. Together they learn and together they evaluate how to learn better. This is one place where you can fill in many of the blank spaces left between the links of the so-called objective tests.

A good comparison can be made between an industrial education teacher and a professional football coach. Each has the responsibility of teaching and assisting individuals to develop their abilities and achievements to their greatest potential. A great football coach must be a great teacher. He must also be an expert in evaluation to work with many players over a long period of time and put together the best possible team. His content is primarily performance and understanding of concepts. His evaluative procedure is primarily conscious observation as an interaction with individuals he knows inside out.

It is not an easy method of evaluation. It is not as quick as making up tests and quizzes. However, it does have a number of ingredients that are so valuable and yet seldom found in student measurement, testing, and grading. Do not get the wrong idea. Observation does not replace other evaluative devices. It is strictly an addition to the program of evaluation and is given special attention here because its potential has not been properly exploited in industrial education.

By conscious observation is meant the planned and studied ability to be alert to what you see, to take on the attitude of consciously recording mentally what your students are doing, how they feel and react. Many people observe but do not really see. Others see what they want to see. A good psychologist sees many things the ordinary person would overlook. He knows what to look for and has developed his senses to a keen edge to observe human behavior. Are you a good observer? As a teacher, now is the time to start developing your observation perception of your students' activities. A number of techniques toward better observation should be helpful to you:

1. You may do a first class job of observing your students' performances, their difficulties, and progress. However, it is difficult to remember all of the mental notes you have taken. These should be recorded on some type of a check sheet for future reference.

2. Your rating devices, often called check lists, rating scales, cumulative records, participation charts, and so on, should be personally developed to suit your needs for conscious observation. They should be planned so you can quickly record judgments and significant details of observed student activity.

3. Some of your rating devices may be quite objective. For example, a rating scale to evaluate student performance in wiring and soldering in electricity, after explanations and demonstrations have been given.

4. You should plan some highly subjective evaluative devices for conscious observation. These will give you an opportunity to record personal judgments and interpretation of student progress, attitudes, behavior change, what he thinks, and how he thinks. THESE ARE HIGHLY IMPORTANT. They support the more factual information received by testing and performance results.

5. Plan your observation to supplement your teaching. Use the insights you discover to assist students all along their paths of learning. You can help change negative attitudes or personal feelings when you have located such difficulties. Observation is one of your best evaluative devices.

Performance Testing and Evaluation

The evaluation of student performance behavior is involved with the action or active part of your teaching. The end result of an industrial education course is an evaluation of how well the students understand and are able to perform. Performance is the vital part of learning, for it reflects most everything your students have studied, practiced, and learned. Students may develop a deep understanding of certain concepts or principles, but until they put them into practice you have no way of knowing how good their performance may be.

To perform and to understand are two different things, and yet they go hand in hand. A performance without understanding is like shaking hands with your dog. When you say, "Shake hands," he automatically lifts his paw. He has learned how, but he does not understand why. There is little place in industrial education for this type of performance. The student needs to understand why he is doing a particular activity as well as to be able to perform the activity.

To this extent, performance evaluation becomes one of the most important parts of the total evaluative process. Any evaluative device should take into account the students' understanding of concepts as he performs. It should also reflect the interest of the teacher in assisting his students to do well. Performance evaluation is an integral part of teaching. It should be used as much as a method of teaching as it is used as an evaluative device. This is why it is necessary that the spirit of performance evaluation be one of teacher-student cooperation and mutual trust. You should never be secretly checking student performance. You should, rather, openly discuss with them what standards they should attempt to attain, what qualities are to be evaluated, and how their activities will be rated.

Another important part of performance evaluation is progress checking. It is often too late to be of value to evaluate a performance after the activity is completed. You should objectively evaluate student activities to see how they are doing and keep them on the right track as they go along. In this way the student is able to take advantage of new suggestions and techniques, avoid possible time wasting efforts due to errors, and make his performance more meaningful. Your "in progress" evaluation also gives you an opportunity to see if a student is performing better now than he has done earlier.

With this background for the evaluation of student performance in industrial education, you now have an opportunity to look at some devices you may plan to develop for your courses.

Performance Evaluation

In order to avoid making quick judgments or generalizations regarding student performance, it is a good idea to develop a check list or rating scale which both you and the student may use as a guide. These may be a "built-in" feature of your presentations and activity sheets or they may be separate check sheets. In either case, a good evaluative instrument will provide for check points throughout the construction of a product, a process, or an experimental activity. To illustrate this point, say you are having students in a junior high school class construct a small mold for vacuum forming and then test their molds in a vacuum forming machine. If you provide ten steps or procedures which they are to follow in their design, construction, and production, you can help them evaluate their progress. These check points may deal with design, workmanship, use of equipment, or any other factors of importance to the process. These will help keep them on the right track, allow for self evaluation, and give you a chance to give further explanations and needed instruction if points are missed.

Keep such rating devices simple and easy for the student to understand. They may require just a check mark for each point or a short statement. In most cases they will be similar to a progress chart, only they apply to just one activity or performance. You should design progress charts in the same manner. They will give you an opportunity to evaluate performance throughout a total course.

Performance Tests

In order to become involved with all aspects of student performance, you may want to plan rating devices in which you use specific performance tests along with subjective judgments. The use of these depends on the goals and desired outcomes of your student activities. Performance tests would usually be used only when you find it necessary to closely evaluate manipulative skills needed to perform a particular job or process. In most cases the teacher develops a planned check list which he notes while observing a student perform. Tests of this type are planned to evaluate how well a student does something; uses drawing instruments, adjusts a carburetor, wires a circuit, pours a casting, or turns threads on a lathe. When it is necessary for a student to develop exacting skills, even in solving a mechanical problem or writing a technical letter, a performance test is a good device for checking a student's ability to apply knowledge.

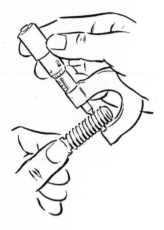

An example of a performance test can be illustrated by checking a student's ability to make certain setups on a metal lathe. Having planned a test in advance, you would observe an individual student perform when you ask him to place the lathe in back gear, adjust the quick change gear box to cut 13 threads per inch, offset the tailstock to turn a particular taper, and so on. As you watched his performance you could easily rate his ability to apply what he has learned. When he has difficulty or his performance is sloppy, you are able to detect it immediately and suggest further study and practice. Observation of performance will usually suffice when a high degree of manipulative skill development is not required. When a high degree of manipulative skill development is called for, make good use of performance testing.

Problem Solving Situations

A good deal of discussion has been given to problem solving as a method of teaching and learning. Some has been devoted to evaluation of the solution of problems. It should be reemphasized here that evaluation of a problem solving activity must be "built in" to the total learning situation. This requires an ongoing process of evaluation in which both student and teacher are involved. It requires a cooperative attitude and an open mind to be receptive to criticism and new ideas. By an ongoing process is meant that evaluation must become a way of thinking on the part of the student. From the identification and definition of a problem to carrying out the final solution, the student must be self evaluating every step. If not, the learning situation is either a waste of time or really not a problem solving activity.

You should not be confused by the term self evaluation during this type of learning. It does not mean that the student works all by himself. It refers, rather, to the introduction of evaluative procedures by the student. He uses all resources he can to assist in his evaluation. These include the teacher, other students, resource persons, and reference materials. A cooperative attitude should be developed toward the best way of doing or working out each step along the way. In the final analysis, you should spend time with each student in reviewing the problem solving activity to see that they understand where improvements could have been made, what concepts they learned, which ones they had trouble with, and if the approach to the problem was satisfactory. Seminars, in which all students participate, are also a useful evaluative device to gain a better perspective of student performance and understanding.

Evaluating the Product

There are times when it may be necessary to rate the results of the student's performance. This is debatable. Much depends on the type of product involved in the student activity. In some cases it would be completely absurd, in others there may be merit to such an evaluation. Agreement must first be reached as to what a student product or project is; how it is defined. In industrial education it is usually thought of as something the student has made, constructed, or produced, either from his own planning or from the plans of the teacher. It also is thought of as something complete in itself, such as a billfold, coffee table, barbecue fork, model house, or other items, from simple to highly complex. It would also have to be concluded that the product is handmade, not machine produced. In other words, if a student pushes a button and a plastic key holder pops out of an injection molding machine, it would appear absurd to evaluate the product in comparison with another student doing the same thing by hand.

Now on to product evaluation. The validity of product evaluation lies in the connection made between student performance and understanding, and the resulting product. If they are closely observed as a total learning process, the evaluation of the product becomes one of the many factors involved. In this case, evaluation becomes similar to that used for problem solving activities. Evaluation of the product then becomes meaningful as it relates to how well the student learned, how his behavior changed, his understanding of mistakes he made, or an improvement in his performance.

Evaluation based on the finished product is risky educational business. An extremely high quality product does not necessarily reflect what happened to the student on his industrial education journey. He may have wasted time and material, broken tools, violated safety rules, failed to understand many concepts, and had considerable help from others. Such things are not reflected in the mirror finish of the product. At least not unless you had a two-way mirror and were also able to evaluate everything the student learned or did not learn along the way.

If you were to visit a department store and evaluate all of the latest model refrigerators, it really would not tell you a thing about all the people involved in their manufacture. What they learned, how they were able to perform, what problems they had, are all hidden behind each refrigerator door. So it is with your students products. Take them into account, but ONLY as they relate to what happened to the student.

More importance may be placed on the results of an activity which may or may not be called a product. In a drawing course for example, certain standards may have been presented and various techniques explained. More emphasis may be placed on evaluating the drawings students produce. A rating scale that has been carefully prepared and checked may serve as an essential device in making such evaluations. It is your responsibility to analyze your particular situation to decide what emphasis should be placed on product evaluation and how it is done.

Types of Written Tests

Written tests have had their share of abuse in teaching and evaluating industrial education students over the years. They are perhaps the most difficult evaluative devices to prepare to correctly serve their intended purpose. The application of written tests is primarily directed to evaluation of your presentations and explanations in regard to student comprehension and understanding. You are actually evaluating the results of this phase of your teaching. Written tests, or paper and pencil tests, are very important in this evaluation of concepts and principles. Performance evaluation does not provide you with all the information you may need to accurately appraise student learning. A student may perform poorly and yet you still might not know what misunderstandings, confusion, or wrong ideas brought about the inadequate performance. You may now see that evaluation of your student's understandings, along with his performance, is something like looking at a pile of materials from which a car is manufactured, and then watch a test driver make the first trial run. Both are an important part of total evaluation.

It should be perfectly clear that your evaluation of understandings not be limited to a unit quiz or a final examination. You will be determining many of these concepts through your questions and discussions during presentations or demonstrations. In other words, your methods of teaching will provide many avenues for you to see if students are "getting the idea" and understand what they are doing. A broad concept of evaluation takes for granted that your evaluation of understanding is a continuous process and an integral part of your teaching. The written test is just an extension of your methods and should be but part of the picture for the evaluation of student understanding. With these ideas in mind, consider some of the following factors in planning all types of written tests:

1. Plan your questions so they do not rely entirely on memorization. Pure recall of facts, names, or dates does not give you as much information as a question that requires a decision between a number of factors to be applied to a specific situation. A decision question tests reflective thinking and concept formation as well as memory.

2. The degree of difficulty of a test should be appropriate for the level of your class. A test in which all students answer every question correctly or all fail does not give you a measure of differences among students. You should strive to attain a cross section of difficulty in your questions.

3. If, in some cases, you are teaching the same material over and over again, you may want to make an "item analysis." Your tests can be improved by continually replacing questions which all students miss with better worded statements or using a different type question. The same would be true for items that all students answer correctly. Some content may be "basic," but many teachers prefer to cover it in a different way for each class and make up new questions for every test.

4. Make your tests clear and easy to understand. Put your wording in the simplest form possible so even the poorest reader can understand. Use short, direct sentences or statements. "Wordy" questions are sometimes misleading. Trick questions should not be used.

5. Cover the content. Avoid using items that include only part of what the students expect the test to cover. A test should emphasize the concepts presented during your instruction. Unimportant items should not be included, they are distracting and misleading to the student. Your tests are just as much learning instruments as they are evaluating devices. Let your students know what to expect. They will learn how to study better and you will receive a more honest evaluation. The length of tests should be just enough to cover the important content.

Selecting Written Tests

Many factors should be considered as you make plans to select tests to evaluate different types of learning. There is no one type of test that will provide you with all you want to know and also be a valuable learning device. The so-called objective tests take a great deal of time to prepare if they are to be as valid and reliable as you can make them. They are less time consuming for the student to take and for the teacher to evaluate. Whether they are more reliable than essay, discussion, or student analysis devices is debatable. In the final analysis, all tests are as objective as your ability to frame questions, the ability of the student to interpret the questions, and your evaluation of responses. In other words, it appears impossible to plan written tests that you could call completely objective.

In industrial education courses, it is seldom possible to secure adequate evaluation of student accomplishment by objective type examination alone. You may find it most desirable to use a number of different types of written tests to enable you to get at all kinds of student growth and development. Each type of test has its advantages and limitations. Every test is of some value under certain conditions. Essay questions provide an opportunity for students to express ideas and organize their understandings. They are time consuming to evaluate and difficult to grade objectively. So it appears that a combination of items on most written tests would provide for a better overall evaluation of student progress. As a good industrial education teacher, you must become knowledgable about all types of evaluative devices to make selections which will best serve you and your students for any learning situations.

A short discussion of a number of written tests will give you an opportunity to analyze their appropriateness for various learning situations. Using items from each or all types may provide the test you desire.

Completion Tests

The planning of completion tests makes it possible for you to secure answers based on recall rather than the student choosing between answers already stated. They require the student to select a word or words that have been omitted from a sentence. When the student places the correct word or words in the proper blank, the statement becomes true.

The advantage of completion tests is that they discourage the possibility of guessing. They save time in teacher grading and reevaluation with the class following the test.

One of the greatest disadvantages of completion tests is that of trying to interpret a word or words that are similar but not exactly what you expected. You have to decide if the answer given is just as good, partially good, or completely wrong. This is usually due to the difficulty of stating completion items so that only one response is correct. A few suggestions for planning completion items along with some examples are as follows:

1. Avoid copying statements directly from student instructional materials or textbooks.
2. Confusion is minimized when you provide only one blank for each sentence.
3. Choose important content words to leave blank and select those words where only one answer will be correct.
4. Whenever possible, place the blank at the end of the sentence.

POOR

The_____ adhesive to use for_____veneer to curved surfaces is_____.

BETTER

The best adhesive to use for laminating veneer to curved surfaces is_____.

Try wording statements so there is no doubt as to one correct answer.

POOR

The correct temperature for pouring aluminum castings is determined by_____.

BETTER

The correct temperature for pouring aluminum castings is determined by using a_____.

Never omit words at the first of a sentence, avoid omitting verbs, and try to place the blank at the end.

POOR

_____of electrons at one_____per second past a given point in a circuit is called an_____.

BETTER

Electron flow at one coulomb per second past a given point in a circuit is known as an_____.

Short Answer or Listing

There are a number of opportunities to devise questions of the short answer or listing type which require students to "think through" a process, make comparisons, or relate a procedure. The advantage of these items are that they determine whether a student is able to explain what he has done or relate what he was supposed to have learned. They also allow for little guessing.

Items such as these are more difficult to evaluate. For example, a student may have correct responses for a procedure but has not placed them in the proper sequence. The number of responses you desire will also aid you and the student. Confusion will exist if you plan an item requesting a list of steps in a process without indicating how many. You may have six steps in mind but the student only lists four. For any listing items, be sure to indicate how many you expect.

Short answer questions give you a good insight into student understanding. Be explicit in your directions. Allow the student to respond with his own understanding of the problem. Short answers are often used for definition of terms, explanation of a process, or the understanding of a concept.

When stating a listing item, be sure the topics are major points, exacting, and conclusive. If you expect them in a proper sequence, indicate your question as such.

The following examples illustrate two ways in which these may be used.

SHORT ANSWER:
Explain the difference between specific adhesion and mechanical adhesion as they relate to the strength of a glue joint.

LISTING:
Indicate five structural differences between hardwoods and softwoods.

Matching Tests

The principle involved in preparing items for matching tests which require more than pure memorization, is to require students to make a choice based on reasoning and understanding. A test that lists a variety of metals and another scrambled list of ores from which they are obtained is a memory test. On the other hand, a list of products made of metals and a list of processes by which the products are most suitably made, measures a degree of ability to choose the right process for the product. The student must understand the process in order to make a correct choice.

It is usually a good idea to have a longer list of choices than words or phrases to match. Then the student is not able to rely on the process of elimination to answer the last few items.

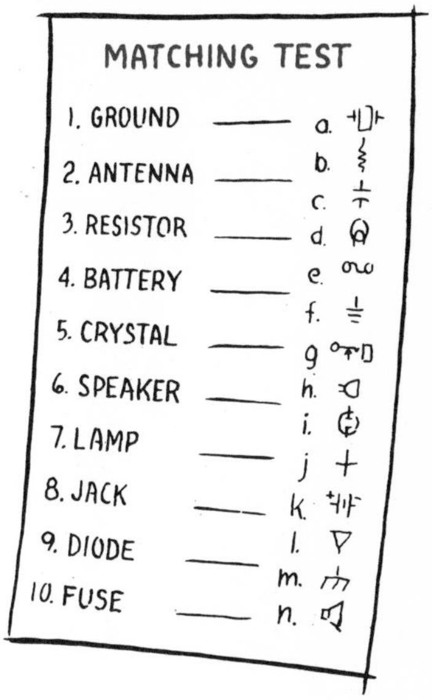

Matching tests provide a number of advantages for industrial education courses, which have so many terms, tools, processes, and equipment. If letters or numbers are used for choices, it saves the time of rewriting all of the terms. Matching tests are also easily adaptable to machine scoring and computerized rating. Since all the terms or words are listed, there is no confusion for the student in writing nomenclature which may be different, but also correct, as in the case of completion items. Remember, matching tests are a good evaluative device if you put a lot of planning "in," so your students may take a lot of understanding "out."

MATCHING TEST

1. GROUND	a.
2. ANTENNA	b.
3. RESISTOR	c.
4. BATTERY	d.
5. CRYSTAL	e.
	f.
6. SPEAKER	g.
7. LAMP	h.
	i.
8. JACK	j.
9. DIODE	k.
	l.
10. FUSE	m.
	n.

Analogy Tests

Another form of written device that may be used for certain evaluations in industrial education courses is the analogy test. An analogy is the inference that if two or more things agree with one another in some respects they will probably agree in others. They can be especially effective in evaluating student understanding of relationship. An analogy requires a student to think through a relationship of ideas, things, or concepts to establish a basis for their relationship to something else. For example, a simple analogy is illustrated below in terms of shape and size relationships.

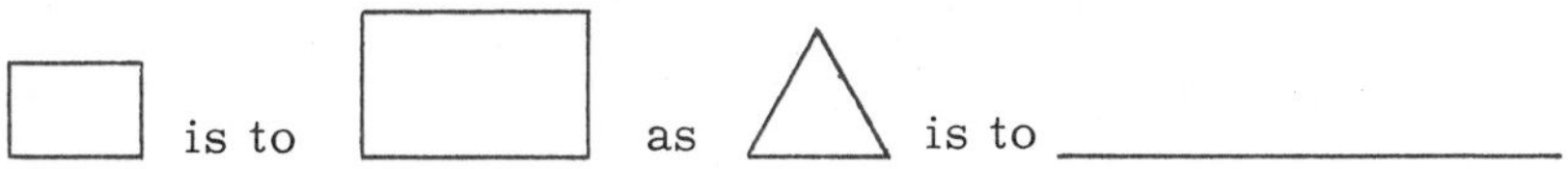

☐ is to ☐ as △ is to _____

In this case, a small square is to a large square as a small triangle is to a large triangle. The relationship of the first three establishes the basis for determining the quality of the fourth. There may be many instances when you would want to use an analogy test item to bring out an understanding of relationships. As is easily seen, relationships of this type can run from simple to highly complex situations.

They have the advantages of being easy to check and also provide an evaluation of some concepts that are difficult to assess with other devices. The major drawback of analogies lies in the ability of the teacher to establish relationships that lead to one direct answer. They should never be used when the relationship of any of the parts is in doubt.

The following may be helpful in preparing test items using the relationship of things or ideas, one to another:

A hacksaw is to metal as a coping saw is to_____.

An electron is to an atom as an element is to a_____.

An inch is to centimeter as a yard is to a_____.

The obvious answers to the above are wood, compound, and meter. As test items the answers should be just as obvious, if the concept has been taught and the student is able to establish the relationship.

Essay Tests

The planning of good essay questions is not as easy as the term sounds. Some teachers may say that essay questions are quick to write but they take a long time to evaluate. This is doubtful if you are really trying to measure a student's ability to apply principles and develop concepts. There are two types of essay questions that cause trouble, those that are too broad and those that request factual answers. Broad questions like, "Explain how copper is made," lead the student into the valley of nowhere. How many answers could be given which would be relatively correct? How could you evaluate such answers and what will the student learn in his attempt.

Essay questions that request factual answers so they may be evaluated objectively might as well be used in a selection type test. They may be misleading and too time consuming when posed as essay questions.

Try the following suggestions when you are preparing essay type questions:

1. Prepare your students for answering essay questions. Teach them how to approach the problem and present principles and understandings.

2. Use essay questions only when they serve a functional need. As with any type test, determine what you want to evaluate and if it is in the form of student explanations of applying principles, use the essay.

3. Use a number of questions when possible, rather than one question to cover everything. Answering may be less tedious and more intelligent for the student.

4. Essay questions should mean the same thing to all students. Do not use questions that could be easily misinterpreted or have double meaning.

5. State your essay question or problem clearly and to the point. Do not leave it up to the student to have to revise your question so he can give an intelligent answer.

6. Make a suggestion as to a time limit for each question.

7. Let your students know how you will evaluate each question and what weight you will place on each.

True-False Tests

A number of statements which are either true or false comprise this evaluative device. The student must indicate which statements are true and which are false by writing a T or F in the appropriate space. Other means are to have students circle or underline the symbols, use a plus or a minus sign, or check yes and no.

Even at its best the true-false test only allows the student a decision between two choices. The test must be skillfully written so your students may benefit by doing some reasoning or reflective thinking. The forced decision of true-false items is illustrated below.

1. T F An alloy is produced by melting and mixing two base metals together.

2. T F Walnut is brown due to the soil conditions in which the tree grew.

A good true-false test is difficult to develop. The first difficulty lies in the fact that the student has a fifty-fifty chance of guessing the right answer. Some statements may be so nearly correct that they could be interpreted as both true and false. By the name of the test you are telling your students they are true or false. However, many educators feel that any statement may be true or false depending on how it is interpreted and in what context it is placed. Take the walnut for example. Soil conditions do determine shades of brown but, on the other hand, it has not been scientifically proven why walnut is brown. Even for the alloy statement, some metals won't melt and mix together. Since half of the test must present false information or concepts, teachers often feel that students may learn misleading ideas when they are not able to make the correct choice. Make your judgments carefully when preparing true-false items.

Multiple-Choice Tests

The term multiple-choice does not accurately describe this type of evaluative device. Most-appropriate-answer would be more fitting if the items are well planned and the possible responses grouped together to require decisions based on the understanding of principles. Students often refer to this type of test as multiple-guess, perhaps because they have been exposed to many items that require the memorization of facts rather than courses of decision.

The main statement used for each item should suggest a course of action, a decision, or a job to be performed. Usually four possible answers are provided. In some instances, one of the answers, or another response, will indicate "all of the above" or "none of the above." You should be careful in your planning to see that a response for one item does not provide an answer for another item. Examples are as follows:

1. The resistance to the flow of electrons in a wire is measured in:
   a. Watts
   b. Ohms
   c. Amperes
   d. Volts
2. When the lips of a twist drill are ground to uneven lengths and the point is off center, it will:
   a. Drill an oblong hole.
   b. Drill a hole larger than the drill size.
   c. Drill a hole smaller than the drill size.
   d. None of the above.

Reports and Research

The evaluation of student activities related to written reports, homework assignments, and research papers plays an important part in these learning situations. In order to make these kinds of learning activities the most meaningful, you should provide your students with some guide lines for preparing their written work and then how you will make evaluations. These guide lines may be in a variety of forms, but they should be consistent so the student knows what to expect. Many of these will grow out of your assignments or planned activity sheets. In other cases, a planned format may be the most valuable device.

Written reports or assignments give the student an opportunity to work independently, develop his own concepts, and establish understandings and relationships. They should be evaluated as objectively as possible and on the basis of the student's ability to explain the meaning of things, develop new ideas, and relate principles to action. You should prepare your students for written work through explanations and class discussions so there will be no doubt as to what you expect and what you will be looking for as you evaluate reports. Caution your students of the worthlessness of copying material from other references. Evaluation is time consuming, but to assist the student in his ability to do technical writing and formulate concepts you should make notes on his paper suggesting improvement or indicating where he got off the track. Take the time to discuss reports with your students, ask questions and help the student evaluate his own work. Just a grade at the top of a report does not give the student much direction concerning the work he has done.

Review Questions - Chapter 9

1. What are the major considerations of a broad concept of evaluation?
2. Explain a number of ways in which you may integrate evaluation into the teaching-learning process.
3. What fundamentals of evaluation lie within the concept, "It is more important what a problem does to a student rather than what a student does to a problem?"
4. In a matching test, why is it good to have a longer list of scrambled choices than problems or words to match?
5. Why is progress evaluation necessary along with testing devices for good evaluation of performance?
6. How are performance and understanding related? Explain why evaluation should take both factors into account.
7. What are the main disadvantages of true-false tests?
8. Why is it so important to evaluate the student's growth and development as an ongoing process of learning?
9. What aspects of learning and evaluation are hindered when the students do not know what to expect from your evaluative devices?
10. Explain what is meant by an analogy test. What aspects of learning would this type of test item attempt to evaluate?
11. Explain what considerations should be made when planning a good essay type test.
12. What is meant when an evaluative device is said to be comprehensive?
13. In a multiple-choice item, what should the main statement suggest to the student?

14. How can you make reports and research evaluation the most meaningful to the student?
15. Explain the important meanings of reliability and validity as they relate to the planning and use of evaluative devices.
16. What are some of the advantages of using short answer or listing items on tests? For what types of learning might they best be selected?
17. Why can it be said that no test is truly objective? In what ways do tests that include both objective and subjective responses aid in evaluation of learning?
18. What factors should be considered when selecting test items for any comprehensive test?
19. In what ways do product evaluation devices serve to evaluate learning? How may they fail?
20. How can the teacher and student best cooperate in evaluating problem solving situations?
21. What is meant by conscious observation? How may observation be validly used as an evaluative technique?

Suggested Activities

1. Prepare a progress chart to be used in a high school industrial education course of your choice. Indicate what performances and understandings you will be evaluating with the chart.
2. Plan the format for a research report to be used in an industrial education class. Explain what evaluative methods you would use to make the results of the report the most meaningful to the student.
3. Outline a performance evaluation device to be used while observing a student going through a process, setting up a piece of equipment, or performing an operation. Include the main topics that you would expect the student to perform and understand.
4. Prepare a written test for a particular unit or topic for an industrial education course. Be careful to select those test items which you feel will best evaluate the types of understandings and knowledge expected to be learned.
5. Make up an essay test which is planned to bring about a knowledge of student understanding of some technical concepts for a unit or topic of your course.

Chapter 10
INSTRUCTIONAL MEDIA

Communication is a vital factor for good instruction in industrial education. The ability of the teacher to effectively communicate with his students in a minumum amount of time is essential to the learning situation. The vast resources of teaching methods that have been discussed provide the starting point for instruction. However, even the best in teaching methods limits the teacher to verbal explanations and demonstrations. It is here that a combination of methodology and instructional media collaborate to provide for the ultimate in learning for the student.

If you have ever tried to describe something technical with words alone, you know the frustrating feeling. You almost feel as if your hands were tied behind your back, helpless in your attempt to communicate effectively. So often you feel if you only had this or that at the moment of presentation, you could get the point across quickly. This is especially true in industrial education courses, where the rapid development of new materials, new processes, new products, and new ways of doing things makes it almost a full time study for the teacher to "keep up." Add to this the preparation of instructional media to assist in explanations and demonstrations, and you have a fantastic job ... the fantastically interesting job of being an industrial education teacher.

Instructional media or communication media are terms used to designate "anything" that can be secured or prepared by the teacher to assist in more effective instruction. They are extensions of the older term, audio-visual materials, which tended to limit the perceptive use of some devices. The use of all kinds of materials and instructional devices is not limited to any particular phase of your teaching. In fact, instructional media would probably be most effectively used in conjunction with all of your teaching methods. As you are making explanations, conducting class discussions, giving demonstrations, leading seminar activities, planning experiments, giving directions, and on and on, there are a multitude of instructional materials which would make these learning situations more vivid, exciting, and educationally meaningful.

The proper use of any instructional media makes your actual teaching job easier, saves time, and helps students learn better. This concept is emphasized again to illustrate two major points. First, it is extremely important for you to make proper selection of instructional media for any teaching situation. Some devices are appropriate and some are not. They do not replace good technical teaching; they make it better when selected and used properly. Second, it is equally as important that you prepare or secure quality media that will do the job intended. So another major part of your industrial teaching responsibility, as if you hadn't had enough already, is to have adequate instructional media ready for each and every presentation.

Selection of Appropriate Media

It is not within the scope of discussion here to indicate any specific devices for use in a particular course. Rather, it is intended to suggest some guide lines for the selection of appropriate instructional media to go along with the various methods of presentation you may be using. Some teachers advocate that selection of instructional media be based on their appeal to the learner's senses, and that they are more meaningful when they appeal to the sense of sight or touch perhaps, rather than just that of hearing. There is no doubt about this advantage and it is discussed later. However, the basis for selection of any media should be directed to its appropriateness to the learning situation at hand. This should be your first consideration. How well will any device assist you in "getting the point across" and help your students learn faster and more comprehensively? It is the prime question you should ask yourself, in a very soft voice, as you contemplate the media you plan to use.

Some general suggestions concerning the selection of instructional media are as follow:

1. ALL STUDENTS SHOULD BE ABLE TO INTERPRET THE MEDIA. There should be no doubt that your students will be able to evaluate and comprehend the device you are using. It should be geared to their level of understanding. If it is too complex, they may be lost. If it is too simple, they may neglect your presentation.

2. THE SIZE IS IMPORTANT. Any physical media should be put into proper perspective. It should be large enough so all can see and interpret, yet small enough to be handled efficiently.

3. MAKE ACCURATE REPRESENTATIONS. Inaccurracy in preparing media may easily lead to student misunderstanding. Secure or prepare materials that are not faulty or misleading. Your devices should be as self-explanatory as possible. If you have to explain what your media is supposed to represent, something is wrong.

4. MAKE PROVISIONS FOR STORAGE. Your instructional media should bé readily available for selected uses. You save time and effort when the materials you need are close at hand and yet protected. An instructional materials storage area is well worth consideration. More specific suggestions for selecting and using instructional devices are discussed with the individual media.

Sense Perception in Learning

Consideration has already been given to the value of appealing to a number of human senses in good instruction. It is only more fitting to concern yourself with sense perception when considering the use of instructional media. The nose, for example, is sometimes neglected, as are the hands and sense of taste. Too often individual senses and combinations of senses are not taken advantage of in fitting your teaching media to the learning situation. Your instructional devices should appeal to the senses that are most suited to the topic of study. If you want your students to really know the odor of brake fluid or burning plastics, let them smell samples. Let them get the feel of the pressure applied to a power drill, the surface finish of a casting, the heat retained in a molded plastic part, or the slipperiness of graphite between their fingers. If you want them to know what is inside a battery, show them a cutaway. If you want them to learn the sound of a dull milling cutter, set it up and let them listen. It's all that simple. Yet, sometimes the teacher does not set the stage for the simplest of things.

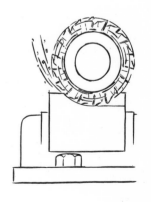

The principle is that you plan your learning activity, select your instructional media, and then apply that media to all the possible perceptive senses of your students. A student may see the brake fluid, watch its action in a hydraulic system, hear an explanation of how it is made, and yet not recognize its odor if a leak is found under the family car. Take enough time in your planning to consider what senses you can best apply to your instructional media.

What to Use and When

The outstanding industrial education teacher develops a knack for knowing what instructional media to use and a sense for timely application. It is doubtful that there is any other area in education that needs and can make better use of such a multitude of instructional materials than industrial education, which represents the whole of American industry. This is why it is perhaps more difficult to give an accurate definition to instructional media in this particular field of study. So, for a broad concept of instructional media, far beyond the audio-visual techniques, it can be assumed that anything the industrial education teacher uses to communicate technical understandings to his students should be included. Alright you say, but that may include billions of things. So it may, perhaps many billions of things. You can't help it, it's your field of teaching and the better you become acquainted with the use of "things" in your field the better educational programs will become.

Take the lonely toothpick for example. Could it be considered a part of instructional media? Certainly, if it were used to "help" you teach something. Go a step further then. Suppose you want to teach how toothpicks are made. Now you are really into something and you need help. That is a difficult story to tell. In order to communicate the necessary concepts and understandings you must rely on numerous instructional aids. A brief list should illustrate the point and give reference to the topic of what to use and when. A box of toothpicks would be essential. Students could then start discussions and questioning. Why are toothpicks made of light colored wood? What kind of wood? Why don't they have a strong wood taste and odor? How are they made so smooth, and many more. To do the best job you would probably need charts and diagrams, overhead transparencies, perhaps pictures or color slides. The machinery from log to toothpick is quite complex. Your selection of "what" to use would depend on those things which would best fit the learning situation for every aspect of the study. These would be planned prior to use. The "when" would depend on the nature of your presentation and discussions. Invariably you will use your teaching media in many ways; to present material, to reinforce concepts, to answer questions, to review, to evaluate, or to correct misunderstandings.

The case of the toothpick answers many multisensory questions. Students hear your explanations of how they are made. They see a variety of media illustrating the manufacturing processes. They smell, for many woods have an undesirable odor. They feel the smooth finish which keeps them from splintering. They taste, and compare with other woods which are undesirable. Best of all they learn, because you have used the necessary media at the right times.

Classification of instructional media is difficult. They may be based on sense perception, ways in which they are used, or more commonly, on the necessary equipment or items themselves. For discussion purposes, classification is made according to groups of materials and any necessary equipment for their use.

Using Instructional Media

It has been found in a number of studies that teachers fail to take advantage of using many instructional devices for two reasons. First, they may not know how to operate various pieces of equipment and tend to shy away from their use. Second, they feel the production of teacher prepared media is too time consuming so they settle for what they have. The reasons are valid, but a good teacher must make up his mind to study the equipment and the preparation of materials much as he does his course content. Actually they go hand in hand. Teachers often indicate that the more they use equipment and prepare materials, the more involved they get with their content and actually learn a great deal by searching for and organizing content to be presented.

Instructional media do not replace the teacher or the various methods he uses. They are supplementary to and part of the overall teaching scheme. The industrial education teacher should take every opportunity to become acquainted with all of the media which will improve his teaching. Some devices have been used for many years, such as the chalkboard and films; others are being developed so fast you can hardly keep up with the literature. The most important media and equipment will be discussed here, with major emphasis being given to the newer instructional devices.

Materials, Equipment, and Products

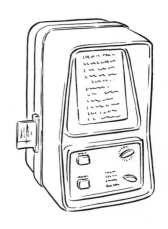

If you haven't already, you might as well start thinking about the most basic devices available to all industrial education teachers, the equipment and products themselves. A well equipped industrial education laboratory will be loaded with industrial materials and products. How better can you teach concepts of castings than to have various types of actual castings available for students to examine, compare, and even test. Your responsibility in this respect is to keep aware of new materials, equipment, and products on the market and request samples whenever possible. You will find that industry is as cooperative in donating sample materials and products as you are in composing an appropriate letter.

Because of the numerous and usually inexpensive materials and products, the plastics industry makes a good illustration. Yet the same is true for most industries. Take any unit you may be teaching in plastics, say blow molding of containers. Get the materials and products in your own hands and in those of your students. Talk about multisensory media. Students can examine the plastics granules, become familiar with their weight, color, odor, and other properties. If each has a blown bottle, examination may give indications of how the mold was made and what trimming operations were required. They may cut the bottles in half and gain an understanding of variations in wall thickness, interior finish, weld lines, and on and on. They may clip a section from the bottle and make a burning test to determine which plastic was used.

The point is made. Learning becomes alive and real under your supervision and direction. Other media would certainly be used to bring about different concepts, but what better choices do you have than the ones described for the desired understandings. It holds true for any technical area. Secure electronic tubes, ceramic products, gasoline engines, wood pulp, powdered metals, tools, stampings, drawings; get them into your students hands and learning begins.

Models, Mock-Ups, and Displays

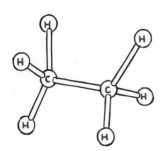

In many cases the actual product or item to be studied is of a size, price, or form that is not readily usable. It's impractical to get a space ship through your classroom door. When such situations arise, a model, simulator, mock-up, or other method of representing the item may be the next best thing. A model of a rocket, a cutaway of a condenser, a demonstration kit or mock-up of a mechanical mechanism, may be just the item needed to assist you in developing the desired principle toward your objective.

These instructional devices have a number of advantages. They are three-dimensional and realistic, giving your students an opportunity to handle, take apart, adjust, and in some cases operate the device. They provide a meaningful contribution to the learning process. Many of these devices are available from suppliers. You may also want to construct your own or have students make models or mock-ups as a part of their learning activities.

Mock-ups usually differ from models in that they have moving parts and simulate the real object. A mock-up of a jet engine, for example, provides an opportunity for the student to get "hands on" experiences which other media may not offer.

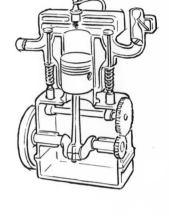

Care should be taken to see that these media really serve the intended purpose. It is educationally wasteful, for example, to have a student or yourself make a wood model of a circular saw blade. The actual blade may be used as an instructional device in its own right. Make your judgments sound and practical. On the other hand, a model of the molecular structure of ethane (C_2H_6) or the flow of electrons through a wire may bring to life some concepts that are difficult for the student to comprehend.

Speaking of models and other simulated devices, they also make wonderful displays for learning. Not the type that were previously discussed to illustrate your course to other students or the public. Here the concern is about learning displays for teaching purposes. They usually consist of sequence units suitably mounted to show how something is manufactured or processed. Such a display may illustrate how a hammer is forged from a beginning slug of steel to the finished product. Another may have the actual parts of an electronic circuit for a radio mounted and wired in full view, so students can follow the conceptual operation of each component.

Your imagination should lead you to at least two conclusions, if not more. The potential for learning by using media of these types is fantastic. Secondly, a teacher's busy mind will be able to create innumerable models, mock-ups, and displays that reinforce student learning beyond belief.

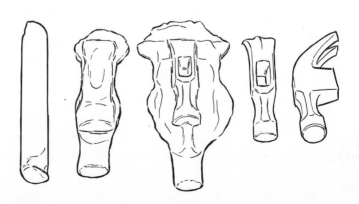

Instructional Hand-Outs

Another form of teaching media that has considerable merit are instructional hand-outs. This topic is limited to printed matter and does not refer to industrial products or supplies. The instructional value of hand-outs is time saving for the student since he does not have to copy material from the chalk board or take notes from a presentation. It also gives you an opportunity to discuss topics, give directions, or present activities while all class members are looking at the same material. This is similar to projection of material on a screen for discussion purposes except that when you throw off the switch the material is gone. In the case of hand-outs, the student may retain the material for future reference and application.

Instructional media of this nature includes those types of things that are not readily available to the student and yet you feel they will aid your instruction and their learning. They would be supplementary to the textbook and may include such materials as diagrams, charts, supply lists, manufacturers' specifications, product drawings, design ideas, sample problems, identification charts and so on. Most of these materials you can prepare yourself from reference sources and industrial literature. These are easily reproduced by ditto or offset lithography for distribution to your students. They are often time consuming to prepare, and to keep on adequate supply requires storage space. So your selection of instructional hand-outs should be determined by the values they will provide to student learning needs and your teaching methods. Avoid copying materials your students may easily obtain.

Another important source of instructional hand-outs is from industry. This is especially true for new materials, supplies, and equipment on the market. Many manufacturers will send you multiple copies of their literature on equipment and products for student use. For example, manufacturers specification sheets on particular adhesives you are using, such as polyvinyl acetate (white glue) or urea formaldehyde, usually provide detailed information on mixing requirements, shelf life, spreading, clamping pressures and times, hot or cold press specifications, and suggestions for a variety of applications. Survey the various industries in your technical field for literature that may enlighten student understanding and make your teaching more comprehensive.

Overhead Projection and Transparencies

The wonderful world of communication and color come alive when you present an exciting topic with the aid of overhead projection and well designed transparencies. Fortunately the overhead projector has had widespread acceptance by teachers, and it is now recommended that there should be one in every classroom. In past years teachers were often reluctant to search all over the building for a projector, sometimes not even being able to locate one at the time they planned their presentation. So they resorted to other instructional media. If you do not have an overhead projector in your teaching area, do all you can to acquire one through your school system.

The advantages of overhead projection for classroom instruction are practically unlimited, but a few of the more important ones are as follow:

1. Most projectors are easy to operate. Usually only three controls are necessary, an on-off switch, a focusing knob, and a device to raise and lower the picture.
2. Since the projector is at the front of the room, you are able to face your class and at the same time have control over the timing and manipulation of material. You are also in a position to use other instructional media, such as models or the chalk board, along with your transparencies, since the room may be fairly well lighted.

3. The transparency on the projector is seen by you in the same context as it is seen on the screen by your students. Therefore you may point to desired areas or write on the transparency in any manner necessary to make "on the spot" communication.

4. You may design and make your own transparencies with fairly limited facilities. Transparencies are commercially available on many topics that may suit your needs. Some are correlated with technical text materials.

With these advantages in mind, you should consider the many techniques that are possible in making overhead projections most meaningful as an aid to your instruction. Using clear plastic film and a wax-based pencil or felt pen, you can write or draw needed diagrams and use them immediately. Using this media is somewhat similar to using the chalkboard except that you are still able to face your class. You have the opportunity to use a pointer with prepared transparencies to further explanations and review questions.

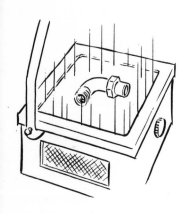

The rate of your presentation can be controlled by covering part of the transparency with a piece of opaque paper and exposing portions as they relate to the points being discussed. Another technique is to use additional transparent sheets as overlays. The base sheet may present the beginning of a process, while the overlays add a step-by-step procedure. An example might be the assembly of an automobile. The base sheet may contain the frame, while each overlay may add the engine, drive system, cooling system, electrical system, brake system, and body.

Plastic parts and three-dimensional objects can be projected by placing them on the glass plate. Opaque objects will provide a silhouette while plastic objects will show color and shape. A plastic drawing template, for example, will project the shapes and lettering in good detail. Even a thin cross-section of a wood species will illustrate important aspects of structure.

A number of special purpose materials may be obtained to simulate motion. By using a polaroid glass spinner over the lens and polarized film attached to the transparency, various types of motion, such as vibrational, reciprocating, and reversing, can be achieved. The polarized sheet material is easily cut to desired shapes and self-adheres to the transparency.

Commercially produced transparencies on many topics for industrial education are available from a number of textbook publishers and commercial concerns. These are available in finished mounted form in black and white or color. Printed "masters" are also available commercially so you can make your own transparencies by any of several reproduction processes. Many transparencies can be made from one master which is printed on heavy paper. Careful selection should be made of commercially prepared transparencies. They are expensive and although some are of high quality, others are not. It is a good idea to order these "on approval" in order to determine if they fit your exact needs. Consider whether they are technically satisfactory, geared to your content and grade level, well mounted, make good use of overlays, and are up to date. The reference section of this text indicates a number of sources of prepared transparencies.

You have virtually unlimited opportunities to prepare your own transparencies. Since you are the one most closely related to your course content, you have the advantage of preparing materials that will correlate with your other teaching media and methods, and really put the desired concepts across. Use them wisely and efficiently. Avoid typing long text material on a transparency. Usually, this may better be presented in hand-out form. Make good use of diagrams, processes, techniques, assemblies, relationships, problems, and other important technical principles. As an industrial education teacher, you are in a unique position of possessing the technical skills to produce outstanding transparencies. Use these skills and knowledge of your content to prepare some of the best instructional media available in assisting your students to learn and enjoy learning.

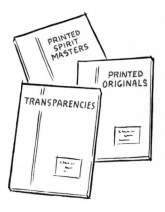

Time and space do not permit in this text a complete discussion of all the materials, devices, and techniques required to prepare transparencies. Many references are available that describe in detail, the various materials and methods that can be used. They also include sources of materials and the operation of reproduction equipment necessary to make your own transparencies. Check with your audio-visual center or reference section of this text for instructions.

Films and Filmstrips

Many films are available for educational purposes, especially industrial education. These have a number of advantages for instructional purposes which are well worth your consideration:

1. Bring the world of industry into your classroom. Films make it possible for your students to view and gain understandings of many concepts and processes not readily attainable through other media. The camera is able to record such things as the melting of metal in a blast furnace, underwater action of a churning propeller, molten glass being pulled into threads, or the inside of a lumber dry kiln in operation; many things that a student could not see or get close to even on a field trip.
2. Change the speed. Films make it possible for your students to see the cutting action of a circular saw blade or the trip hammer of a forge in slow motion. They may also view the cur-

ing action of a cast plastic material or the forming of corrosion on a metal surface in a few minutes, which may take many hours in reality. This is an outstanding feature of this media.

3. Films provide for "seeing in action" combined with sound. Sensory perception is a contributing factor for better learning in situations such as the total process of forming a steel ingot into a roll of sheet steel. The many individual actions may be seen close-up while the sounds of each operation are recorded.

4. Interest and understanding are promising. Students who lack interest, for a variety of reasons, may be highly motivated by an exciting and colorful film presentation. Films also overcome the inability of some students who lack reading skills or audio terminology to relate both to the actions and sounds of a clearly presented film.

5. Films provide a common denominator for discussion and understanding. After viewing a film, all of your students have a common experience from which they may base questions and further discussions. In many instances the film is but the beginning of a total learning experience.

6. Films may be used quite effectively as an introduction to manipulative skill development. Equipment and machine operation can be viewed close-up by your students. This is similar to an actual demonstration except that the camera can pick up details that are difficult to see, record operations in slow motion, and review important procedures. The techniques for good arc welding, for example, may be closely observed by a large group without any possible damage to their eyes. This is a vast improvement over the demonstration where conditions are crowded and students are unable to view the procedure at close range.

7. Safety first. Many films which illustrate safe working conditions around machinery and equipment prove to be one of the best media to bring about a safety consciousness on the part of students.

These advantages of using films in industrial education courses overshadow the few disadvantages. Of main concern is the selection of films for the intended purpose. No two teachers present their courses in the same manner, so it is difficult to locate films which express principles or show processes exactly the way you would like. Secondly, many films contain sections dealing directly with your topic of study but often include considerable material that is completely unrelated. Therefore it is requisite that you preview any film for possible future use. A very disappointing learning situation can prevail by showing an unpreviewed film.

Films for industrial education can be secured from many universtiy film-loan services, local and state film collections, and from commercial rental libraries. However, some of the best films can be obtained from manufacturing companies and industrial organizations. Many concerns have educational programs of their own and produce films for their own courses. Most of these are readily available to you upon request. Additional film sources are listed in the reference section of this text and in many textbooks in technical areas.

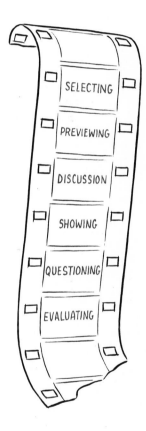

Like most other instructional media, class readiness is a necessity if student learning from a film is to be effective. Prepare your class. Lead them from what they know to what they might expect to learn from the film. Discuss new terms that will come up during viewing. List questions which the film may partially answer. Discuss any technical details you feel students may misunderstand prior to showing. After viewing, follow up with a discussion of the topic. Use questions which were developed prior to showing the film. You may also want to evaluate understandings by an oral check or written report. Do your best to make your student viewing-listening experience interesting and meaningful.

The same principles for films apply to the use of filmstrips. Most filmstrips consist of a series of still pictures on 35mm film. They are often in sequential order for presenting processes and various concepts. In general, they are little more than the incorporation of many drawings, pictures, graphs, or charts into one compact length of film. They are easily handled, shipped, and stored. A main advantage in their use is that you can spend as much time as necessary on any one frame, or reverse to other frames for review. They are especially valuable in presenting factual information in visual form and showing "how" things are done for group instruction in skills. Filmstrips are relatively inexpensive to purchase. Many may be obtained on loan. Hand viewers, some of which are battery lighted, serve both to preview filmstrips and allow students to study material much as they would a book or other printed material.

Filmstrip projectors are also available which include sound. These are table top units for individual as well as for small group viewing. They contain the viewer, record player, and controls for stopping and starting at any time. Notes may be taken or problems solved at desired intervals. Study carrels or reference centers make ideal settings for this equipment. Use the reference section of this text for filmstrip sources.

Slides

Industrial education teachers find the use of slides highly advantageous in the presentation of many forms of technical content, from complicated industrial processes to simple laboratory sequences. Slides used in the classroom are the same as photos you might take with your 35mm camera during a camping trip. You send the film to a dealer to be processed and mounted.

Many slides may be purchased, usually in color, from commercial suppliers. Special series prepared for industrial education courses are available. However, the content is often presented in a manner that does not meet your particular needs. Therefore, the industrial education teacher is in a good position to make use of his talents in picture taking using a 35mm camera and color film. A number of references provide complete instructions for use of the camera and suggestions for good picture taking. In many cases you may "fill in" a commercial set of slides with your own pictures to tell the story the way you like. Teachers are turning more and more to making their own slides and purchasing or exchanging sets with other teachers by having duplicates made. Proceeding along this line, in a relatively short period of time you can have ideal materials for a number of your instructional units.

Using slides as a teaching method is similar to the use of filmstrips as previously discussed. In most cases they will be presented in a sequence to study a process, such as the operation of a laminating press or an automotive compression tester. Slide presentations do not replace the demonstration, but they may be supplementary to it or provide for an introduction to a

process. They also make a valuable means for presenting processes or operations on pieces of equipment you do not have available in your laboratory. It is important that you provide the verbal captions or notations to help your students interpret what they see as the viewing proceeds.

Slides are also one of the best media you have to bring close-ups of actual objects into the classroom for group study. For example, close-up stands are available on which to mount your camera to make copy work slides. Photographic copies of mechanical drawings, electronic circuits, or welding flame adjustments are often advantageous for discussion purposes. Another extremely valuable tool is an attachment to mount your camera on a microscope. This process is used extensively for preparing slides to study material structure. Microscopic view of paper fibers, metal fractures, or the cellular structure of wood make it possible for you to reinforce concepts with your whole class, when these are projected on the screen. You can imagine what an aid this is to student learning when you compare it with trying to help individual students see what you want them to see through a microscope.

For individual study, continuous slide projection cabinets are just "what the doctor ordered." These table model units have rear projection on a small screen. Up to 100 different slides can be sequenced into a slide cartridge and can be operated automatically, by remote control, or by hand. Recording cartridges of magnetic tape make it possible for you to record explanations to coincide with each slide. Up-dating is fast and inexpensive, since you can replace slides at any time in the cartridge and rerecord your message. This is an advantage over films and filmstrips.

Your students can individually view selected topics for investigation, advanced study, or review. Since the image is so bright on the small screen, there is no need to dim the lights or pull shades.

Single Concept Films

Because of the trend toward the use of 8mm film and projection equipment in schools, the topic of single concept films is treated separately. The use of 8mm projectors is now dominating the field once devoted to 16mm equipment, so your attention should be directed toward the use of this media in industrial education courses. Two major factors have dominated the trend. First, the development of 8mm optical and magnetic sound track projectors, and secondly, the introduction of the cartridge type film system.

The cartridge film system does not require threading or rewinding. A rigid plastic case encloses the film. The film cartridge is placed in an opening in the projector and the switch turned on. Some of the newer projectors automatically prefocus, turn off automatically, rewind the film, and leave the machine ready for the next showing. The projectors are either the rear view type, for individual or small group viewing, or the front screen models, which can be used by individuals or large groups. An individual student may also use a headset for listening which eliminates distracting sound.

The principle of single concept films is to present a single idea, process, or understanding in which the use of motion enhances the learning potential. Most of these film cartridges will present viewing from thirty seconds to four minutes. Students may run the film over and over again to reinforce the concept presented. For self-instruction, the lightweight projector may be located anywhere in the classroom or laboratory. A single concept in welding or sealing plastic film, for example, can be viewed "on the spot" by a student and put to immediate practice or application. A number of movie-pack films are available on a variety of topics for use in industrial education classes.

Instructional Television

The use of the television media in industrial education has unlimited potential. Equipment for many applications is available and in use in schools and universities throughout the country. It primarily consists of a multipurpose television camera, television sets to receive the picture for student viewing, and equipment for recording video-tape. This equipment will provide for the major uses of television in teaching-learning situations. In many cases commercial television and educational television programming, from cities or communities, has been found to be of little value in industrial education. Therefore, your concern here should be toward the advantageous use of television as it applies to better understanding of your technical content.

The major advantages of using television techniques over other instructional media are:

1. To present to your students material that is not available in any other form.
2. To provide image magnification in motion for demonstration purposes.
3. To show on-the-spot operations that may be dangerous to a group of students.
4. To tape record presentations or demonstrations to be available for reuse.

You should become acquainted with the techniques necessary to prepare closed-circuit television presentations and video-tape presentations.

Closed-Circuit Presentations

A portable television camera can be mounted on a stand or tripod in a fixed position over a demonstration table or piece of equipment. The camera may be connected by a wire circuit to a single receiver or a number of receivers depending upon the size of your class. As you make the demonstration, say the adjustment of the needle valve on a carburetor, you would give your explanations in the usual manner as your class watches the television screen. Your performance is similar to a typical demonstration except that you should rehearse your performance so that the materials and your hands will be in positions that are the most readily viewable. Discussions, questioning, and explanations provide for student comprehension while your performance is being televised.

When the camera is in a fixed position, it is necessary for you to adjust the controls and lighting at the demonstration table. However, when you plan a demonstration at a piece of equipment, perhaps a spot welding machine, it requires another person to operate the mobile camera equipment. If you work with another industrial education teacher, you can both perfect your techniques in zooming in for close-ups or backing away when greater viewing area is required. Try your hand at being a television performer, technician, producer, and camerman. You will enjoy the experiences and your students will enjoy the opportunity to learn through another important media.

Video-Tape Recording

When you have made closed-circuit television presentations and they are completed, you must repeat the performance for any other situation. If there are demonstrations or presentations that you would like to preserve for repeated viewing, you can record them on video-tape. You may also want to record student performance for immediate replay and evaluation for your class. In any case, you can convert the camera to a video-tape recorder and record a demonstration from a fixed position. You may also record from a mobile unit as another person demonstrates. Another method is to use a completely portable, battery operated unit that can be carried and operated by one person. The video-tape recorder is carried on a shoulder strap while you hold the camera by hand. This makes it possible for you to record pictures on trips to an industry or of activities in your laboratory. These recorders will generally give you about 30 minutes of televising, and you may record audio explanations at a later time while replaying the video-tape. This equipment may also be run on regular AC current. Now you can be a wandering educational television camerman, recording important content for your classes. The video-tapes may be stored or erased and reused at any time.

Programmed Learning Materials

Programmed instruction is a process used by students to direct their learning toward instructional objectives with a minimum of teacher intervention and direction. The programs are prepared by the teacher in a number of forms depending on the equipment to be used. They may range from simple paper-pencil programs to complicated equipment programs.

The two major types of programs, adapted to any equipment, are the "linear" and the "directive." The linear program provides all students with the same stimuli in the same sequence. The student responds in written form, by punching a hole in a card, or by pressing a button. In general, the student completes a series of items in an order of progression. The items are usually problems or questions which lead from one to another, with cues to help the student supply the correct answer. Of main concern, is the reinforcement of correct answers. The program can then be self-contained, and the student may carry the learning experience from start to finish without outside help. A few items from a linear program dealing with the thermoforming of plastics may better illustrate how a program is planned:

1. Thermoformed plastics are heated to soften in order to attain the desired shape of the object. Therefore, thermoformed plastics become _____ when heated.

soft 2. Soft plastics are then heated and _____ into shape.

stretched 3. Once heated and stretched, the plastic will retain its shape when cooled. The final step in thermoforming is _____.

cooling 4. Heating, stretching, and cooling are the three characteristics of the _____ process.

thermoforming 5. The basic methods of thermoforming are mechanical forming, vacuum forming, and blow forming which are the results of _____ , _____ , and _____ of plastics.

heating, 6. (continued to the desired instructional objec-
stretching,   tive)
cooling

The directive program is more complicated to plan, for it provides for individual differences and an opportunity to relearn through supplementary information. Rather than linear, it is a branched program in which the student may make choices for correct answers. If his answers are correct, he goes right on. If not, his response is evaluated and he may be detoured for further information or another assignment, with explanations of why he was wrong. He may be told to return to a previous question and then try again or he may be told to go to another source of information before trying again. Many paths may be planned for directing the learning process.

The better directive programs are planned for mechanical or computerized machines which make use of 35mm slides, cartridge tapes, printed diagrams, and films. So much equipment is on the market in the form of teaching machines and programmed instruction units, that you should refer to the references for detailed descriptions. In any event, you should study the possibility of adapting portions of your content that are the most suitable, for programmed instruction. Directions are given with each piece of equipment for planning your own programs.

The main advantages of programmed instruction lie in the speed of student learning and comprehension and teacher freedom to assist other students in learning activities.

Field Trips and Industrial Resources

Probably one of the most valuable learning experiences for industrial education students are field trips or industrial visitations. Here is the opportunity for your students to become involved with the "real" things and people of our industrial society. They can observe, hear, smell, and touch the world of work, from minute operations to automated manufacturing and production. They can talk with workers, designers, researchers, technicians, management, and employment personnel to gain an understanding of the attitudes and practices of those employed in industry. Make use of any opportunities you have to bring student understanding to its best, when a visit to industry will most adequately meet the objective.

Some considerations you should make in planning field trips to an industry, museum, research laboratory, or any other community resource are as follow:

1. Will the trip provide the kinds of experiences that fit the units of study your students are pursuing?

2. Do you have enough time to complete the trip and see what you want students to observe?

3. Can you make adequate arrangements for a tour, preplan the trip with a guide, and prepare your students for what they should learn?

4. Have you made arrangements with your school administration, transportation procedures, finances, and parent permission?

5. Have you discussed proper dress, behavior standards, and safety precautions with your class?

6. Would a class discussion of pertinent questions to be asked during the visit be profitable?

7. Will the trip be profitable for your age level students?

8. In what way may you plan an evaluation of the experiences gained from the visit?

It is a good idea for you to explore your industrial community resource. Get to know people. Find out what phases of any industry may be the most meaningful for your class to visit. As you develop a relationship with local industrial personnel, you open the door for inviting specialists in many fields to serve as guest speakers for your class. Safety, research, job opportunities, educational requirements, and manufacturing are typical of the subjects on which guest speakers may provide up-to-date information. Many companies will be glad to have their personnel give demonstrations or show films and slides to your classes. In some instances they will gladly supply products or partially processed materials to each member of your class for further study.

Your industrial resources are actually too numerous to mention. Look into every aspect of the industrial situation in your community and state dealing with your technical area. As discussed earlier, you can also use the mail to bring many industrial resources into your classroom.

Review Questions - Chapter 10

1. What are the unique advantages of using overhead transparencies in presenting technical content in industrial education courses?
2. Explain what possible resources are available to you from industry that would serve as good instructional media.
3. What are the two main considerations you should give to the proper use of instructional media?
4. Present a good explanation of the relationship between sense perception and learning in regard to instructional media.
5. Why do teachers often fail to take advantage of the variety of instructional media available to them?
6. Explain how materials and products from industry can be effectively utilized for instruction purposes in industrial education courses. Give some examples.
7. Why have 8mm films and film cartridges become so extensively used in recent years?
8. What is meant by a single concept film? Illustrate some purposes they could serve in industrial education courses.
9. Describe the two classifications of programmed instruction materials.
10. When do models and mock-ups serve most effectively as instructional media?
11. Explain the different ways overhead transparencies can be secured and prepared for instructional purposes.
12. What major plans should you make when considering a field trip to industry? In what ways should you prepare your class?
13. What disadvantages are there to the selection and use of

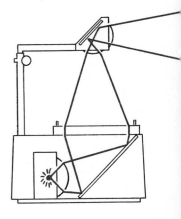

films which you should carefully consider?

14. Why are the many forms of instructional hand-outs a valuable aid in good teaching? In what ways do they serve better than other media?

15. How may guest speakers aid in supplementing your instruction?

16. Explain how colored slides which you prepare yourself can be most effectively used for technical presentations. Also explain their use for technical presentations. Also explain their use for individual student study.

17. What are the advantages of filmstrips in instruction? In what ways are slide cartridges easier to keep up to date than filmstrips?

18. What are the three major components of a basic television system suitable for school use?

19. What advantages do television systems have over other instructional media that make them very practical for use in industrial education courses?

20. How can displays be used for instructional purposes as opposed to showing what is being done in your course?

Suggested Activities

1. Write to distributors of printed programmed instruction materials to secure literature about available programs. Prepare a report for the class on how you might use these materials in an industrial education course.

2. Make a list of instructional hand-outs that would be advantageous for you to prepare for your technical course. Be sure to make your selection as to their value over other instructional media.

3. Prepare a paper and pencil programmed instruction device for a carefully selected topic for your technical course. Try it out with some students and make necessary revisions after evaluating the results.

4. Make an outline for a colored slide sequence for your selected course to assist in teaching a process, or concept. Briefly indicate the content to be included in each slide.

5. Secure a piece of material or product dealing with your course content and explain how it could be effectively used to aid in teaching if placed in the hands of each student in your class.

6. Prepare a chart or diagram that could be used to present a concept or understanding along with your explanation. Select a topic which would make the best use of this media for your instruction.

7. Plan a selected list of films and filmstrips you could use for a junior high school class in your technical area.

Chapter 11
PERSONALITY
AND LEADERSHIP

A good teacher is an excellent leader. He must work closely with his students to establish the best learning relationship possible. The teacher's personality is directly related to many of the outcomes of leadership and instruction.

You should continuously analyze your own personality to assist you in better developing your personal qualities of leadership so essential to good instruction. Many of your abilities to promote student understanding go far beyond your knowledge of subject matter and technical skills in making presentations and giving demonstrations. These are very important, but your ability to inspire student growth and development are equally important. They usually grow out of your leadership ability and personal relationships with your students. So take enough time to sum up some of the more important aspects of personality and leadership. Also give yourself the opportunity to relate these qualities to your students, for it is important that they develop their own personalities and leadership qualities under your supervision. Just as sawdust rubs off your arms during a demonstration, so do kindness and character.

Establishing Student Confidence

The setting for industrial education students is an ideal spot for you to assist them in attaining some of the "fringe benefits" of learning. They may be called fringe benefits because most

do not show up in course content or reference materials; yet they are some of the most valuable outcomes of learning that your students may have an opportunity to acquire. A few of the personal qualities on the part of your students that should develop out of a good teacher-student relationship are kindness, consideration for the rights of others, courtesy, friendliness, helpfulness to others, loyalty, cheerfulness, good work habits, and a love for learning. Education has really begun when you watch students leave at the end of a course with these qualities "glowing." Too often you hear teachers say that they just can't do anything with some students. This is probably true in a number of instances. However, it may be worth recalling the first line of an old poem that states, "Someone said that it couldn't be done, but he with a chuckle replied, that maybe it couldn't, but he would be one, who wouldn't say so till he tried."

One of the first steps is to establish student confidence. Let them know that you like them, enjoy being with them, and trust them. Take a positive attitude toward working with your students. They will find out very quickly the ground rules that will prevail during your course, both those that you state in "do's" and "don'ts", and those that you air perceptively through your personality and attitude. Try a little friendliness - - it goes a long way in securing their confidence in you. If you still find you have a belligerent student, an uninterested student, or an irresponsible student, just don't let anyone say you haven't tried. You can't win every ball game, but you can maintain a mighty high batting average.

A number of things have been noted as to what you may do to establish student confidence. Now take a look at some things that can keep you from stumbling along the way:

1. Do not directly criticize students, especially in front of others in the class. Use your criticism discretely. Show them what they have done well and then suggest improvements, both in their work and their attitude.

2. Do not talk down to your students from the "altar" of the intellectual chief. Talk with them, at their level and in their language.

3. Do not be sarcastic or make jokes about students. You have noted a number of comments concerning making learning fun and telling stories. Just be sure that your jokes do not insult or cause student contempt. Tell your jokes in such a way that they will like you more for having had a chance to laugh a little.

4. Do not tell students they are dumb. They may go out of their way to prove it to you.

5. Do not teach to the slowest segment of your class, nor to the highest. Try to interweave your instruction throughout all levels so each will be challenged but not frustrated.

6. Do not bluff your students or try to cover up mistakes.

Let them know when you are wrong and explain why. They may admit their mistakes more easily to you and realize that you are just as human as they are.

7. Do not continually introduce interesting but unrelated information. Your mountain climbing adventures may hold their attention for an hour, but save that for a mountain climbing course. Much time can be wasted when students can get you interested in discussing your favorite hobby or sport.

Student confidence in you as a teacher depends heavily upon your ability to develop personal qualities of leadership. It is necessary for you to take an inventory of your leadership potential for all aspects of the teaching-learning environment.

Developing Leadership Ability

In order to bring about the types of attitudes and behavior you desire in your students, they must have confidence in you. The factors previously mentioned will certainly help you gain this confidence. The next step is to provide outstanding leadership for them to use as an example in learning and to continuously direct your students toward the desired goals. Leadership does not mean that you teach your students to be good followers. On the contrary, it means that you provide direction and assistance so they may take on the qualities of leadership that you display during their learning.

Leadership in instruction indicates that you must always maintain control of the learning situation. This role requires you to direct the shifting of attention to the focal point of learning. At times you will be the focal point or center of attention during learning. However, you will find it desirable to shift the attention of your class to an individual student at various times. A student may be giving an explanation, showing an operation, or answering a question. In other instances, you may wish to

shift attention to some instructional media. You may want to shift attention rapidly from one student to another as a group discussion is taking place. This is leadership. The ability to direct student attention to the planned points of learning at the proper time is of prime concern to prevent distractions. In short, you should keep the flow of learning directed toward immediate goals and discourage any wandering into areas of unprofitable understanding.

The confidence you develop in your own abilities to direct learning and the student confidence vested in you are the tangible results of leadership in instruction. This combination prevails upon your students the willingness to accept your guidance toward the established goal of learning. Your leadership ability will then take on meaningful proportions, for leadership also saturates the climate of motivation and evaluation. As your students are delving into the daily process of learning and doing, your motivational insights will continue to be a guiding force. You will learn to detect expressions and actions which indicate student doubt, misunderstanding, or frustration. A tug on the arm, a word of encouragement, a suggestion for direction, are but a few ways to remove the stumbling block and guide your student along the road to learning.

The same concept applies to evaluation, for actually the insights of your students to gain an ability for self-evaluation is your goal. A good teacher, with profound leadership ability, is able to shift the emphasis of evaluation more and more to student responsibility. The questions you ask, the comments you make, and the relationships you establish with your students provide them with insights to evaluate their own behavior. Isn't this the real concept of evaluation, for the student to be able to evaluate himself in order to better learn or perform throughout life?

You need to take stock of your leadership abilities, develop them as a conscious act of mind and action, and also see to it that your students have an opportunity to acquire these same traits.

Personality - Success Factor in Teaching

Assuming a person has acquired the knowledge, resources in organization and methodology, and other factors necessary to become a good industrial education teacher, there is little doubt that PERSONALITY is the real factor for success. Not just to be a good teacher, rather the personality traits that constitute the possibilities of greatness in teaching. Many personality traits are quite subtle. Others stand out like the sun coming over the horizon. The personality of a teacher is closely related to his leadership ability, for they rely upon one another in all phases of instruction.

The many factors dealing with teaching success and personality are highly complex. A simplified expression you often hear is that a particular teacher has a pleasing personality. However, the depths of teacher-student relationships control how you act with your whole class and with individual students. The outstanding teacher appears to be on the same "wave length" as his students. That is, he has intuitive or thought out comprehension of how his students will react to his actions. This opens up the whole realm of communication, the expressive media by which students become highly receptive to the learning process.

There are a number of personal characteristics which involve personality that are noticeable in the makeup of outstanding teachers. The overriding characteristic that permeates the atmosphere is the term so seldom found in educational literature - - love. You hear teachers say they love to teach. What do they really mean? They love to work with students? They love the contents that they study so diligently? Or, do they mean to say they love students and, therefore, enjoy helping them learn. Perhaps the latter would prevail if you could note the personality characteristics of, say, the fifty greatest teachers in history. At any rate, it is an intangible ingredient that far overshadows a dislike for students, even if they may be annoying and uninterested.

Evaluate your own personality. Like any other form of change in behavior, it is possible to make changes in your personality. Many weaknesses in personality may be overcome by concentrated efforts to improve. As you read through the following list of personality characteristics, think through your own personality and analyze and evaluate yourself. Also make reference to how you can instill these same qualities in your students:

VOICE: A great deal of your personality is revealed through your voice. You should be able to speak with command and yet be soft and gentle as the situation requires. You can speak with authority and yet listen with understanding to your students.

KINDNESS: A wonderful personality trait to which students most readily respond. Whet the appetites of your students by showing them you like them and enjoy being with them. More than likely they will be kind to you, and go out of their way not to let you down in their behavior or responsibilities for learning.

PATIENCE: The combined efforts of teaching and learning are a slow process. At many times it is easy to become discouraged when things do not go the way you planned. A patient teacher will take stock of the situation and persist in making improvements. It is that quality of personality that allows you to keep calm and keep right on trying even though adverse conditions stand in your way.

SINCERITY: An outstanding teacher must possess this important personality trait. Students readily sense a lack of sincerity and their respect for you as a teacher and person is easily lost. When you show you are sincerely interested in their welfare, in helping them learn, and in "sticking" with them through personal difficulties, you illustrate a vital force in human relationships.

STUDENT INTEREST: When students attain the feeling that you are more interested in them than you are in yourself or the content, you are on your way to good teaching. A student becomes important to himself when you show how interested you are in his accomplishments. Also, your interest in your students brings out your understanding of their individual differences, and a knowledge of individual differences is a requisite for good instruction.

ENTHUSIASM: Although it has been discussed extensively, enthusiasm is contagious. It rubs off and makes for enthusiastic students. Be vigorous, active, exciting, and illustrate what fun it is to learn, by your actions and your teaching.

FAIRNESS: Students react in a positive manner when they feel they are being treated fairly. When you spend more time with some students, others may feel left out or that you do not care. Be fair in giving yourself to your students and in evaluating their progress.

Your personality is a tremendous success factor in teaching. It also plays an important part in the moral and social development of your students. When you display many fine personality traits, students will tend to emulate your actions. Most students would like to be like one of the finest persons they have ever known. Make yourself one of those persons. It will make you a better industrial education teacher, a leader in the classroom, and a leader in your profession.

Review Questions - Chapter 11

1. Explain how personality and leadership are related. How does one depend upon the other?
2. What factors of leadership contribute to more successful teaching?
3. How does instructional leadership play an important part in the teaching-learning situation?
4. What can you do as a teacher to help establish student confidence? Why are these so important for learning potential?
5. In what ways does patience help to improve learning and student confidence?
6. How can your personality play an important role in the social and moral development of your students?
7. How is it possible for you to evaluate your own personality and attempt to make desirable changes which would improve the teaching-learning climate?
8. Attempt to explain why the term love is seldom found in educational literature, and yet is often expressed verbally by teachers.
9. What is meant by the expression that an industrial education teacher is a multidirectional person?

Suggested Activities

1. Prepare a list of terms that you feel would describe the ultimate personality traits of an outstanding industrial education teacher. Use these for a class discussion.
2. Plan a student learning activity which would place some degree of leadership ability on members of the class. Use a class discussion to evaluate what qualities your students feel they have gained from the experience.
3. Make a report, using reference materials, on how instructional leadership may be displayed in teaching using the following topics; demonstrations, instructional media, student activity sheets, and student evaluation.
4. Prepare a bulletin board display that focuses attention on teacher-student cooperation in establishing student confidence. Use pictures and words to emphasize your major points.
5. Obtain an appointment with a personnel director of a local industry. Discuss with him the success factors he looks for pertaining to personality and leadership in persons he interviews for employment. Make a report to the class on your findings.

Acknowledgments

This book has been the culmination of the author's experiences with students and faculty at high schools and universities over a period of many years. Informal discussions with graduate students, professional colleagues, and teachers in the field have provided many insights into the problems of teaching industrial education.

Although it is impossible to personally acknowledge all assistance received, the author would like to sincerely thank Mr. Norman Delventhal, Dr. Clois Kicklighter, Dr. H. James Rokusek, Mr. Alfred Roth, Mr. Harry Smith, and Mr. John Weeks of the Department of Industrial Education, Eastern Michigan University. Their suggestions were invaluable in preparing many sections of the manuscript.

The author would like to pay a special debt of gratitude to Mr. Kendall Starkweather, Industrial Education Teacher, Huron High School, Ann Arbor, Michigan, for many exciting hours of conversation dealing with the traits of an outstanding industrial education teacher.

A special word of appreciation is given to Professor Raymond LaBounty, Chairman of the Department of Industrial Education, Eastern Michigan University, for his confidence and encouragement in preparing this book.

Final credit is given the author's wife, Marjorie, for her untiring assistance in reading the manuscript and making suggestions which would encourage any student to excell in teaching industrial education courses.

SELECTED

REFERENCES

"A Guide for Equipping Industrial Arts Facilities." Washington, D. C.: American Industrial Arts Association.

"A Guide to Improving Instruction in Industrial Arts." Washington, D. C.: American Vocational Association, 1968.

American Council on Industrial Arts Teacher Education. "Problems and Issues in Industrial Arts Teacher Education." Fifth Yearbook. Bloomington, Illinois: McKnight and McKnight Publishing Company, 1956.

American Council on Industrial Arts Teacher Education. "Readings in Education." Sixth Yearbook. Bloomington, Illinois: McKnight and McKnight Publishing Company, 1957.

American Council on Industrial Arts Teacher Education. "Planning Industrial Arts Facilities." Eighth Yearbook. Bloomington, Illinois: McKnight and McKnight Publishing Company, 1959.

American Council on Industrial Arts Teacher Education. "Research in Industrial Arts Education." Ninth Yearbook. Bloomington, Illinois: McKnight and McKnight Publishing Company, 1960.

American Council on Industrial Arts Teacher Education. "Essentials of Preservice Preparation." Eleventh Yearbook. Bloomington, Illinois: McKnight and McKnight Publishing Company, 1962.

American Council on Industrial Arts Teacher Education. "Action and Thought in Industrial Arts Education." Twelfth Yearbook. Bloomington, Illinois: McKnight and McKnight Publishing Company, 1963.

American Council on Industrial Arts Teacher Education. "Approaches and Procedures in Industrial Arts." Fourteenth Yearbook. Bloomington, Illinois: McKnight and McKnight Publishing Company, 1965.

American Council on Industrial Arts Teacher Education. "Evaluation Guidelines." Sixteenth Yearbook. Bloomington, Illinois: McKnight and McKnight Publishing Company, 1967.

American Council on Industrial Arts Teacher Education. "A Historical Perspective of Industry." Seventeenth Yearbook. Bloomington, Illinois: McKnight and McKnight Publishing Company, 1968.

American Council on Industrial Arts Teacher Education. "Industrial Technology Education." Eighteenth Yearbook. Bloomington, Illinois: McKnight and McKnight Publishing Company, 1969.

Arnstein, George E. "Automation - The New Industrial Revolution." Washington, D. C.: American Industrial Arts Association, Bulletin No. 4, 1964.

Bakamis, William A. "Improving Instruction in Industrial Arts." Milwaukee, Wisconsin: Bruce Publishing Company, 1966.

Billett, Roy O., Maley, Donald, and Hammond, James J. "The Unit Method." Washington, D. C.: American Industrial Arts Association, 1960.

Bloom, Benjamin S., et. al. "Taxonomy of Educational Objectives. Handbook I: Cognitive Domain." New York: McKay Company, 1964.

Brown, James W., Lewis, Richard B., and Harcleroad, Fred F. "A-V Instruction - Materials and Methods." New York: McGraw-Hill Book Company, 1969.

Callahan, Sterling G. "Successful Teaching in Secondary Schools." Chicago, Illinois: Scott, Foresman and Company, 1966.

Cochran, Leslie H. "Innovative Programs in Industrial Education." Bloomington, Illinois: McKnight and McKnight Publishing Company, 1970.

"Color Dynamics." (grade schools, high schools, colleges) Pittsburgh, Pennsylvania: Pittsburgh Plate Glass Company.

"Developing Human Potential Through Industrial Arts." Washington, D. C.: American Industrial Arts Association.

DeVore, Paul W. "Technology - An Intellectual Discipline." Washington, D. C.: American Industrial Arts Association, Bulletin No. 5, 1964.

Diamond, Robert M. (ed.) "A Guide to Instructional Television." New York: McGraw-Hill Publishing Company, 1964.

"Educator's Guide to Free Films." Randolph, Wisconsin: Educators' Progress Service, Revised Annually.

"Educator's Index to Free Materials." Randolph, Wisconsin: Educators' Progress Service, Revised Annually.

"Energy in Color" Pittsburgh, Pennsylvania: Pittsburgh Plate Glass Company.

Erickson, Carlton W. H. "Fundamentals of Teaching with Audiovisual Technology." New York: The Macmillan Company, 1965.

Ericson, Emanuel E., and Seefeld, Kermit. "Teaching the Industrial Arts." Peoria, Illinois: Charles A. Bennett Company, 1960.

Evans, Rubert N. "Foundations of Vocational Education." Columbus, Ohio: Charles E. Merrill Publishing Company, 1971.

Fisher, B. M. "Industrial Education: American Ideals and Institutions." Madison: University of Wisconsin Press, 1967.

Friese, John F. and Williams, William A. "Course Making in Industrial Education." Peoria, Illinois: Charles A. Bennett Company, 1966.

"Frontiers in Industrial Arts Education." Washington, D. C.: American Industrial Arts Association.

Gerbracht, Carl and Robinson, Frank. "Understanding America's Industries." Bloomington, Illinois: McKnight and McKnight Publishing Company, 1971.

Giachino, W. J. and Gallington, R. O. "Course Construction in Industrial Arts, Vocational and Technical Education." Chicago: American Technical Society.

Glaser, Robert W. (ed.) "Teaching Machines and Programmed Learning, II, Data and Directions." Washington, D. C.: Department of Audiovisual Instruction, National Education Association, 1965.

Green, John A. "Teacher Made Tests." New York: Harper and Row Publishers, 1963.

Hartsell, Horace C. and Vennendaal, W. L. "Overhead Projection." Buffalo, New York: Henry Stewart, Inc., for American Optical Co., 1965.

Hawken, William R. "Copying Methods Manual." Chicago: American Library Association, 1966.

Hornbake, R. Lee. "New Horizons in Industrial Arts." Washington, D. C.: American Industrial Arts Association.

"Industrial Arts Education: Purposes, Program, Facilities, Supervision." American Council of Industrial Arts Supervisors, Washington, D. C.: American Industrial Arts Association.

Keane, George R. "Teaching Industry Through Production." Washington, D. C.: American Industrial Arts Association.

Kempt, Jerrold E. "Planning and Producing Audiovisual Materials." San Francisco: Chandler Publishing Company, 1965.

Kigin, Denis J. "Teacher Liability in School Shop Accidents." Ann Arbor, Michigan: Prakken Publications, Inc., 1963.

Krathwahl, David R., et al. "Taxonomy of Educational Objectives. Handbook II: Affective Domain." New York: McKay Company, 1964.

Mager, Robert F. "Preparing Instructional Objectives." Palo Alto, California: Fearon Publishers, 1962.

Maley, Donald. "Contemporary Methods of Teaching Industrial Arts." Bulletin No. 8, American Industrial Arts Association.

"Man-Society-Technology." The 1970 AIAA National Convention Proceedings, Washington, D. C.: American Industrial Arts Association.

McAshan, H. H. "Writing Behavioral Objectives." New York: Harper and Row Publishers, 1970.

Micheels, W. J. and Karnes, M. R. "Measuring Educational Achievement." New York: McGraw-Hill Book Company, 1950.

Miller, Rex and Smalley, Lee H. "Selected Readings for Industrial Arts." Bloomington, Illinois: McKnight and McKnight Publishing Company, 1963.

Minor, Ed. "Simplified Techniques for Preparing Visual Materials." New York: McGraw-Hill Publishing Company, 1968.

"Modern School Shop Planning." 6th ed. revised. Ann Arbor, Michigan: Prakken Publications, Inc., 1971.

"New Concepts in Industrial Arts." Washington, D. C.: American Industrial Arts Association.

Olson, Delmar W. "Industrial Arts and Technology." Englewood Cliffs, New Jersey: Prentice-Hall, Inc., 1963.

Osborne, Burl N. "Industrial Arts is for Human Beings." Washington, D. C.: American Industrial Arts Association.

Robertson, Van H., Editor, "Guidelines for the Seventies - T and I Division, American Vocational Association." Chicago: American Technical Society.

Rose, Homer C. "The Instructor and His Job." Chicago: American Technical Society.

Ruley, Morris J. "Leadership Through Supervision in Industrial Education." Bloomington, Illinois: McKnight and McKnight Publishing Company, 1971.

Sanders, Norris M. "Classroom Questions, What Kinds?" New York: Harper and Row Publishers, 1966.

Silvius, G. Harold and Bohn, Ralph C. "Organizing Course Materials." Bloomington, Illinois: McKnight and McKnight Publishing Company, 1961.

Silvius, G. Harold and Curry, Estell H. "Managing Multiple Activities in Industrial Education." Bloomington, Illinois: McKnight and McKnight Publishing Company, 1971.

Silvius, G. Harold and Curry, Estell H. "Teaching Successfully in Industrial Education." Bloomington, Illinois: McKnight and McKnight Publishing Company, 1967.

Smith, Lavon B. and Maddox, Marion B. "Elements of American Industry." Bloomington, Illinois: McKnight and McKnight Publishing Company, 1966.

Taylor, Calvin W. and Williams, Frank E. "Instructional Media and Creativity." New York: John Wiley and Sons, Inc., 1966.

"Technician Education Yearbook." 5th edition, Ann Arbor, Michigan: Prakken Publications, Inc., 1971.

Wilbur, Gordon O. and Pendered, Norman C. "Industrial Arts in General Education." Scranton, Pennsylvania: International Textbook Company, 1967.

Wood, Dorothy W. "Test Construction." Columbus, Ohio: Charles E. Merrill Books, Inc., 1960.

INDEX